Systems Methodology in Social Science Research

Frontiers in Systems Research:

Implications for the social sciences
Volume 2

The objective of the series is to develop a rich resource of advanced literature devoted to the implications of systems research for the social sciences. The series includes monographs and collections of articles suitable for graduate students and researchers in academia and business, including rewritten Ph. D. dissertations. No undergraduate textbooks or reference books are included. Quality, originality, and relevance with respect to the objectives of the series will be used as primary criteria for accepting submitted manuscripts.

Systems Methodology in Social Science Research

Recent Developments

EDITED BY
ROGER CAVALLO

Springer Science+Business Media, LLC

Library of Congress Cataloging in Publication Data

Main entry under title:
Systems methodology in social science research.

(Frontiers in systems research; v. 2)
Bibliography; p.
Includes index.
1. Social science research. 2. Social sciences–
Mathematical models. 3. System analysis. I. Cavallo,
Roger E. II. Series.
H62.S98 300′.72 80–25707
AACR1

ISBN 978-90-481-5808-9 ISBN 978-94-017-3204-8 (eBook)
DOI 10.1007/978-94-017-3204-8

Originally published by Kluwer•Nijhoff Publishing in 1982

For Lydia,
the loveliest little lady

CONTENTS

Preface xi

I FOUNDATIONS 1

1 Science, Systems Methodology, and the "Interplay between Nature and Ourselves"
Roger Cavallo 3

2 On the Use of Structured Methodologies in General Systems Research
Franz Pichler 18

3 Fuzzy Systems Theory: A Framework for the Analysis of Humanistic Systems
Lotfi A. Zadeh 25

II GENERAL METHODOLOGICAL APPROACHES: THE THEORY/ DATA INTERFACE IN SOCIAL SYSTEMS INVESTIGATION 43

4 The Methodology of *Q*-Analysis Applied to Social Systems
Ronald H. Atkin 45

5 General Systems Modeling of Conflict within Nations
Roger Cavallo and *Eduard Ziegenhagen* 75

6 Theory of Measurement of Impacts and Interactions in Systems
Thomas L. Saaty 94

III SYSTEMS-BASED TOOLS FOR SOCIAL SCIENCE RESEARCH 111

7 Social Networks and Multilevel Structure: System Description and Simulations
Stein Bråten, Eivind Jahren, and *Arild Jansen* 113

8 Hypotheses behind the Sociological Interview: Test and Reformulation
Johannes van der Zouwen 142

9 A System Concept and Its Impact on Multiobjective Decision Models
Christer Carlsson 158

Name Index 185

Subject Index 189

List of Contributors 193

PREFACE

From the beginning, the systems research movement has shown a high potential for offering a conceptual framework for the understanding of social systems. Much of this potential has been realized, but a major gap remains with regard to operational investigative aids. Developments of the last ten years with a methodological orientation and emphasis seem finally to be filling this gap. The purpose of this book is to describe the most advanced of these developments and to make them available to a wider audience. The emphasis is on developments that are primarily oriented toward interaction with expertise in the social sciences and that thus hold the most promise for social systems investigation. In particular, attempts have been made to provide substantiation and illustration of three main points: (1) the common motivation and essential integrability that systems research provides for developments and considerations along a very broad spectrum of interests; (2) the very diverse nature of the types and forms of considerations that may be meaningfully integrated; and (3) the operational and usable nature that developments in systems methodology represent for research in the social sciences.

The book is divided into three parts with a generally increasing degree of specificity. The first part (Chapters 1, 2, and 3) deals with foundational issues associated with modeling and methodology as areas worthy of study in their own right. The issues involved are associated with the ability to make general, theoretical, or abstract statements about "real-world" situations—issues that are relevant to the investigation of any complex humanistic system. The second part (Chapters 4, 5, and 6) describes and illustrates general methodological approaches that elaborate useful mechanisms for dealing with the theory/data interface in social system investigations. In addition to considering the methodological approaches, these chapters describe and elaborate particular utilizations of them.

The third and last part of the book (Chapters 7, 8, and 9) deals with approaches that are similar in spirit to those of Part II but that have a slightly more focused orientation. Each chapter represents a realization of systems theoretic ideas in the development of tools for important problems associated with the investigation of social and sociotechnical systems: the description and simulation of social networks, the sociological interview, and decision making in the face of complex situations that involve multiple objectives.

I am pleased to have been able to serve as editor for such a collection of concepts, procedures, and approaches to the investigation of complex humanistic systems, and I have no doubt that the ideas represented in this book—and further developments associated with them—will have a major effect on our ability to deal effectively with major current and future problems.

I FOUNDATIONS

The main purpose of this book is to link appropriate and useful abstractions and problem-solving tools with phenomena or problems of "real-world" interest. The form of the abstractions considered ranges somewhat broadly, as does the form of the problems. The coherence among chapters results from the recognition that the *process* through which abstractions and abstract tools become matched with problems of concern is of major importance. The conceptual construct used in all of the considerations is that of system, and the book presents advances in the attempt to build a problem-solving language for dealing and interacting with complexity.

Any attempt to construct this language must deal with the interrelation between the processes of abstraction and interpretation. The first part of this book contains considerations on a more general level of the manner in which the study of systems and the study and development of methods for systems research serve to reunite symbolic production with problems of empirical and technological concern.

1 SCIENCE, SYSTEMS METHODOLOGY, AND THE "INTERPLAY BETWEEN NATURE AND OURSELVES"

Roger Cavallo

SCIENCE AS INTERPLAY

In *Physics and Philosophy* the Nobel Prize-winning physicist, Werner Heisenberg (1968), argues that one of the most important features of modern physics has been to instigate the development of a "really different attitude toward the problem of reality." This different attitude is one that forces a reexamination of the tenability of a view toward science based on interpretations of the Cartesian partition into God-World-I—a partition that Heisenberg claims "simplifies in a dangerous way the basis for further reasoning." He argues that, although Descartes' position recognized the "indisputable necessity" of the connections between the three parts, developments in the centuries following Descartes saw the polarity between the "*res cogitans*" and the "*res extensa*" emphasized, and science "concentrated its interest on the *res extensa.*" Under the influence of this partition, which "has penetrated deeply into the human mind during the three centuries following Descartes," a view toward science has developed that places an emphasis on science and scientists as separate from nature and whose goal is to "simply describe and explain nature."

Heisenberg continues that the lessons learned from modern physics support the viewpoint that although "we can indeed proceed without mentioning ourselves as individuals, we cannot disregard the fact that natural science is formed by men. Natural science does not simply describe and explain nature; it is part of the interplay between nature and ourselves."

The study and characterization of this interplay, and thus of science—or more accurately, of aspects of science that are independent of the substantive (content) part of nature that is involved in any particular interplay—is, in my opinion, the major characteristic of systems research. Systems methodology, in turn, is the orientation of systems research that is primarily devoted to the study and development of methods used in facilitating this interplay.

Clearly, the previous orientation of science, which Heisenberg insists must change, has proven to be extremely useful and fruitful for many classes of problems. Indications are strong, however, that many problems and situations demanding of our attention today are not susceptible to this orientation. Various arguments can be made to describe the difficulties—for example, arguments in terms of complexity, large-scaleness, or uniqueness of major problems—but all of these arguments seem to hinge at one point or another on problems associated with the involvement of a human element.

Now Heisenberg clearly intends that the "human element" cannot be excluded even from the study of the simplest of systems. For our purposes, however, we need recognize only that social systems and the problems associated with them must surely involve this element. Indeed, a somewhat more extreme evaluation of the effect of the orientation that Heisenberg writes about is offered by another Nobel Prize-winning scientist, the economist F.A. Hayek, who claims that this orientation "has contributed scarcely anything to our understanding of social phenomena" (Hayek, 1955).

Based on Heisenberg's position that science is a part of the interplay between nature and ourselves and on the evolution of investigative modes that attempt specifically to contribute something to our understanding of social phenomena, this volume describes, elaborates, and illustrates major approaches that have recently been developed in systems research. While the diversity of foci represented in the articles of this volume is obvious, a common theme, characterized well by Heisenberg's emphasis on science as "interplay" and by his stress on the importance of *language,* is clearly evident.

Heisenberg relates the importance of language directly to the interplay that represents a mediation between the abstract and formal statements that have characterized "scientific explanation" and the reality, facts, or observations they purport to explain. He argues:

> One of the most important features of the development and the analysis of modern physics is the experience that the concepts of natural language, vaguely defined as they are, seem to be more stable in the expansion of

> knowledge than the precise terms of scientific language, derived as an idealization from only limited groups of phenomena. This is in fact not surprising since the concepts of natural language are formed by the immediate connection with reality; they represent reality. . . . On the other hand, the scientific concepts are idealizations . . . but through the process of idealization and precise definition the immediate connection with reality is lost. [Heisenberg, 1958, p. 200]

It is interesting to consider fuzzy systems theory (see Chapter 3) from Heisenberg's perspective on language. In the development of fuzzy systems theory, Zadeh and others are attempting to define the foundation of what Zadeh expects will be the "systems theory of the future—the systems theory that will be applicable to the analysis of humanistic systems." Zadeh believes that the distinguishing characteristic of this systems theory will be "a conceptual framework for dealing with a key aspect of humanistic systems, namely, the pervasive fuzziness of almost all phenomena which are associated with their external as well as internal behavior." In his paper Zadeh presents the basic ideas underlying fuzzy systems theory and provides an access to the associated literature, all of which represent—as, for example, with work dealing with "deductive verbal models of organizations"—current attempts to come to grips with Heisenberg's perception.

The rationale for Zadeh's description of fuzzy systems theory—the attempt to develop formal, abstract, and specifiable modes of investigation that constitute useful working algorithms and that, at the same time, not only do not deny but agree to and accept the inherent vagueness of "reality"—is a dominant feature of systems methodology. This feature promises to compensate for tendencies that have developed in the last two centuries that seem to be a deterrent to progress on many important problems of complexity. These tendencies are all associated with attempts to separate and presume independence of two fundamental processes involved in scientific "interplay"—those of *description* and of *interpretation* (see Löfgren, 1978; Cavallo, 1979*b*). Expression of this separation can be found, for example, in the large amount of mathematical research that appears to be conducted as if no justification for this research is needed other than that which comes from the formal linguistic structure in which the research takes place. On the other hand, we find that in the context of specific complex problems, researchers who are interested in these problems must either accept abstract structures as given and adapt their problems to them or resort to the attitude that formal and mathematical considerations have no other than isolated and occasional utility and bearing on "reality." Successful elaboration of the complementary nature of description and interpretation holds significant promise for amelioration of these extremes and for the development of abstract structures that are nontrivially adaptable to real problems of concern.

One of the most widely used descriptive characteristics of the systems perspective is the attempt to elaborate a useful paradigm that is alternative or complementary to that of reductionism. This paradigmatic aspect is also relevant to Heisenberg's characterization as he expresses the shift from dogmatic or metaphysical realism—based on assumptions or emphases of descriptions of reality that are independent of scientists, observers, or modelers—to that of pragmatic realism and an inclusion in emphasis of the *purposes* for which descriptions are used and models are built. From the methodological perspective, the increasing consideration of modeling activity motivates a shift from emphasis on given modeling forms to development of conceptual frameworks that accommodate—and within which to integrate—methodologically pluralistic approaches. The basic goal is to allow problems to determine solution procedures rather than the other way around.

THE MODELING PROCESS AS A FOCUS OF STUDY

Any given attempt to model "real-world" phenomena must deal with the integrated use of *measurement, description,* and *interpretation.* The problem represents basic epistemological issues and is one that must be confronted by any scientist who attempts to deal with the interface between specific, unique facts and general, abstract statements.

This chapter gives the background and description of an operational framework and approach to this question, which is based on systems concepts. Chapter 5 illustrates the use of the framework through an investigation of intranational conflict phenomena. The framework that is described and utilized here characterizes a general approach to systems problem formation and solution termed the General Systems Problem Solver (GSPS). GSPS makes use of the dynamic and constructive tension between context-dependent and context-independent considerations to provide a means for knowledge extension, and thus an increased potential for interaction with any complex system.

The method used by GSPS is to develop a comprehensive definition and classification of problems that arise in systems investigation. The classification provides a framework that can be used to structure modeling activity. At the same time, the classification is flexible enough to allow the situations themselves to be viewed as a complex and dynamic set of interacting problems (see Ackoff, 1972). While problem classification constitutes the structural framework of GSPS, dynamic and process-oriented aspects are supplied by developing general systems problem-solving tools and by associating them with defined problems (or by associating already-existing tools with the overall framework). The tools in general represent abstraction through mathematical, computational, and logical facility, while the systems concepts and problem definitions provide the bridge to phenomenological significance.

The problems dealt with by GSPS—those associated with the theory/data interface—are especially important for the social sciences where lack of experimental control forces the social scientist to accept high levels of interaction—and thus of complexity—as a basic aspect of any investigation. This high interaction is expressed from two different perspectives. One expression is from the perspective of the "uninvolved" observer, as interaction among parts of objects of investigation. A more important expression, however, which somewhat subsumes the first, relates to interaction between observer and observed and stems from the perspective of the impossibility of strict separation between the scientist (or observer) and that which the scientist is observing or studying. The importance of this latter aspect is intensified as social scientists increasingly consider more comprehensive objects of investigation, necessitating integration of studies by different researchers, and thus increasing the need for interscientist communication. Modern research in the social sciences is consequently much more aware of the importance of the *language* that an investigative body utilizes.

This awareness has led to the development of an epistemological orientation that contradicts a more traditional perspective represented, for example, by Popper's claim with respect to cognition that "the study of the products is vastly more important than the study of the production" (Popper, 1972). What is termed a "drastic departure" from Popper's position is directly dealt with, from the point of view of the philosophy of science, by Rescher (1973*a*, 1973*b*, 1977) in the development and articulation of a systems-theoretic approach to the theory of knowledge. Equally relevant to this book, however, is the fact that evidence of this departure also appears increasingly in the attempts of social scientists to articulate their recognition of the importance, if not primacy, of *methodology* for social science research (see, for example, Lazarsfeld, 1972; Cortés, Przeworski, and Sprague, 1974; Przeworski and Teune, 1970).

However, one of the difficulties arising from methodological considerations made from within the context of specific problem areas within the social sciences is that context-dependent aspects associated with the specific problems have a constraining influence on the development of *generally* useful investigative aids—that is, on the development of a generally acceptable methodological language. Indeed, the importance of methodological approaches that will be useful for the study of broad classes of social systems provides the main rationale for much of general systems research and systems methodology.

One of the important goals of general systems research is to provide a transdisciplinary language that is primarily devoted to the investigation of real-world objects (see Cavallo, 1979*a*) and that extracts, develops, and operationalizes conceptions and methods common to all investigative areas and disciplines (see, for example, Atkin, 1974; Gaines, 1978; Mesarovic and Takahara, 1975; Wymore, 1976; Zeigler, 1976). A characteristic feature of efforts in this area is a belief that *modeling activity* is important enough to warrant a concentration and major focus of attention.

From a methodological point of view, the framework utilized here is important because of its development in two directions. One concentration is on the development of individual *tools* that are useful in the investigation of complex systems and that provide working methods that extend information-processing capabilities. The focus of this development is not on automated abstract tools as such but rather on tools that can be used interactively by investigators in various contexts to take advantage of the best aspects of both the context-dependent *expertise* and the context-independent processing aspects, as well as their interaction. As opposed to automated problem solving, the approach may more accurately be termed one of "intuition-amplification" (see Cavallo and Pichler, 1979).

The second line of development is on the *integration* of these methods into an overall characterization of the modeling process, which is based on ideational constructs relevant to any system. With respect to models and modeling, the orientation operationalizes Rescher's recognition that the importance of (the modeling) *process*—represented by the continual construction and development of models in an organized fashion—is equal to or exceeds that of any particular models that result. This orientation is especially relevant in the context of the investigation of *complex systems,* and its utility may in fact serve as an extensive definition of complexity.

SYSTEMS DEFINITIONS AND SYSTEMS PROBLEMS

It is fruitful to consider that modeling activity represents a formal and conscious expression of general knowledge and learning processes relevant to all levels of intelligence, but one that is suited and adapted to learning situations of a generally complex nature (see Chapter 2). In this regard, a useful interpretation of such processes is one that recognizes the subject-object interaction as dynamically posing *problems,* while attempts at solution provide the mechanism to add increased knowledge to the overall situation (see Pask, 1969; Piaget, 1970). This problem-solving aspect of the epistemological process contributes the dynamic and constructive component to the organized general systems investigative framework, GSPS. A more thorough explication of the framework GSPS can be found in Cavallo and Klir (1978) and Cavallo (1979*a*). The intention here is only to provide enough background to indicate the manner in which general systems ideas can be operationalized and then indicate the manner in which GSPS can be useful for systems investigations. Thus, only the major components of GSPS, as a framework that may be used in the investigation of any system and, in particular, in the investigation of complex social or sociotechnical systems, are presented here, and the reader is referred to the other literature for details.

GSPS on the one hand defines a framework intended to allow the integration of systems concepts that can relate to any context-dependent consideration. Some of these concepts are, for example, behavior, state, structure, organization, and regulation. All of these concepts appear regularly in the investigation of social and political systems, but their meaning and significance are often undeveloped or clouded because of inconsistent and differing usage. An important aspect of GSPS is that all of the integrated systems concepts are given a methodologically unambiguous general expression enabling their clear and communicable use.

On the other hand, GSPS is defined in a manner that allows access to and integration of results from abstract and more context-independent languages, such as mathematics, computer science, and philosophy. The main object of the development of GSPS can then be described in terms of developing the most suitable matching of results from these abstract disciplines with general systems problems and thus—through amplification and interpretation—with real-world situations that the general systems problems represent.

The core around which GSPS is defined is a hierarchy of epistemological levels of systems. The hierarchy expresses the idea of various levels at which an investigator may "know" an object of investigation. Some such hierarchy is implicit in any investigation or modeling endeavor, though lack of a clear statement often causes unnecessary difficulties through effects associated with hidden assumptions. The statement of the hierarchy utilized in GSPS is one that has been developed in detail over the last decade or so from the original conception presented by Klir (see Klir, 1969). Interestingly, this hierarchy closely parallels the description by Singer (1971) of "levels of knowledge" in any scientific investigation. Singer describes these levels as existential knowledge, correlational knowledge, and explanatory knowledge. There is an obvious correspondence between these types of knowledge and the somewhat more detailed levels of systems definitions as used in GSPS and described in the sequel. The correspondence may be visualized as in Figure 1.1. In Figure 1.1 solid arrows relate systems definitions to levels of knowledge described by Singer and dotted arrows point to systems definitions used in GSPS for which the corresponding level of knowledge is not explicitly described by Singer. The importance of the hierarchy of system levels stems both from its comprehensive nature—starting with an object of investigation and working through levels to metasystems of various orders—and from the fact that the systems at each level are given a clear and general description. This description facilitates both the statement of all assump-

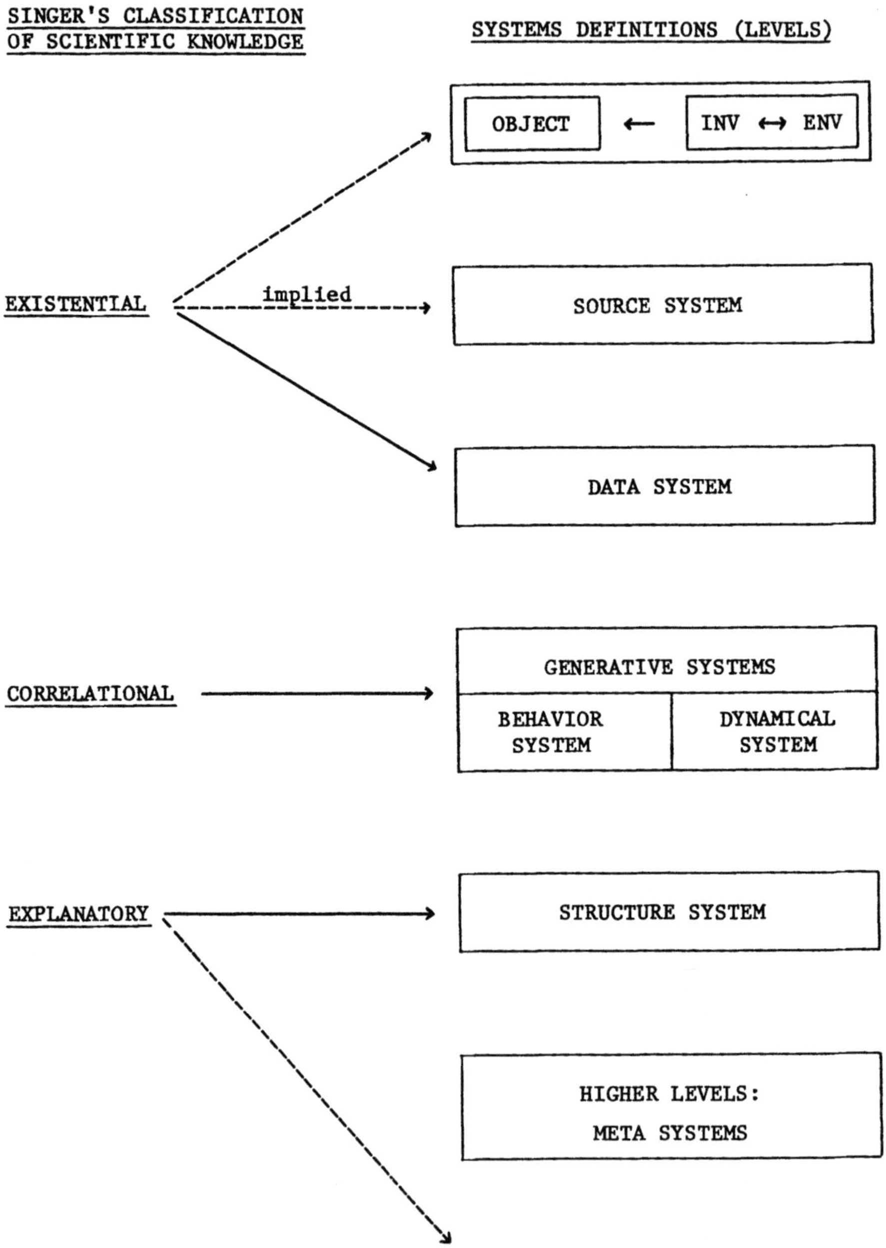

Figure 1.1. Parallel between Singer's Classification and the Basic Hierarchy of Systems Definitions Used in GSPS

tions involved and the development of working methods that are useful for dealing with systems problems.

The General Systems Problem Solver does not deal with context as such, and for this reason the consideration of "object" is restricted as prior to general systems methodology. We thus consider as premethodological in any investigation three primitive notions:

1. An *investigator,* who may be an individual or group;
2. An *environment* with which the investigator interacts;
3. An *object* that the investigator, through interaction with the environment, constructs, conceives of, or formulates from some purpose.

These notions all depend on context and do not allow for significant general methodological statement. Methodological significance arises when a set of *basic attributes* (representable by abstract variables) is chosen by the investigator to represent the object for some specific purpose. In addition to the basic attributes, *supporting attributes* (variables) are chosen with respect to which measurement of the basic attributes may take place. Thus there is associated with each basic and supporting attribute (variable) a set of *appearances* (abstract states) that the investigator is able to recognize. Systems defined at this level are called *source systems* since they represent all possible data that may be gathered. The use of data that do not have representation in the source system essentially constitutes a redefinition of the system. At the level of source system, however, there is no knowledge regarding the ways in which variables are related.

Systems at the next highest level are referred to as *data systems* and are given by defining a function that associates an appearance of each basic variable with each compound element consisting of values of the supporting attributes. This function is defined either through observation and measurement (data gathering) or as a desirable data system (in the case of design).

Systems at the next highest level are called *generative systems* and are defined by giving an overall relation among the basic attributes, which does not depend on particular appearances of the supporting attributes. Systems at this level include dynamical descriptions, state-defined systems, behavior systems, and so forth. The term *generative* is used in recognition of the fact that a properly determined relation can be used to generate the data of the data system. Linear regression represents a method of generative system determination that is particularly popular in the social sciences. GSPS would usually utilize more general procedures.

Systems at the next epistemological level—*structure systems*—are defined by determining sets of interacting (coupled) subsystems that, under specified procedures, adequately represent the overall system (relation) or some chosen properties of it. "Adequate representation" would ideally be "exact" representation, but some information will in general be lost in refinement (or decomposition). In these cases the degree of approximation of various structure systems should be specified and incorporated as part of the representation. A type of structure system determination that has received heavy utilization in political and social systems investigations is one that results from the use of factor analysis. Again, however, because of assumptions associated with the technique, alternate and more general methods are preferable for utilization within GSPS. Cavallo and Klir (1979*a*, 1979*b*) give a comprehensive specification of the technical aspects as well as fully implementable algorithms that determine a structure system from a given overall (generative or behavior) system. In terms of Singer's classification of knowledge, such procedures provide a means to determine "explanations" for observed relational patterns.

Details regarding the different systems levels, as well as the application of the concepts to various contexts, can be found in the literature cited in the references, as well as in Chapter 5. A simple example is described here, however, to illustrate the operational orientation of GSPS.

Table 1.1 represents a simple general system at the level of behavior system defined on a group of workers all performing the same task. Three attributes—sex, age, and performance—are measured, where *sex* is either female or male; *age* is young or old; *performance* is poor or good. As represented in the general format of Table 1.1, sex, age, and performance are ν_1, ν_2, and ν_3, respectively; the values female, young, poor performance are represented by 0, and male, old, good performance by 1. Table 1.1 lists the eight possible states, along with the probability of occurrence of each state, where the probabilities are based on frequencies.

Figure 1.2 gives a representation of the structures in the set of all meaningful decompositions of a system of three variables (these structures are often referred to as "models" in social science literature). Cavallo and Klir (1979*b*) describe the algorithms by which these structure systems may be recursively generated for

Table 1.1. An Overall Behavior System Defined by Observation of the Attributes, Sex (ν_1), Age (ν_2), and Performance (ν_3)

ν_1	ν_2	ν_3	*Prob.*	ν_1	ν_2	ν_3	*Prob.*
0	0	0	0	1	0	0	0
0	0	1	0	1	0	1	.1
0	1	0	.2	1	1	0	0
0	1	1	.3	1	1	1	.4

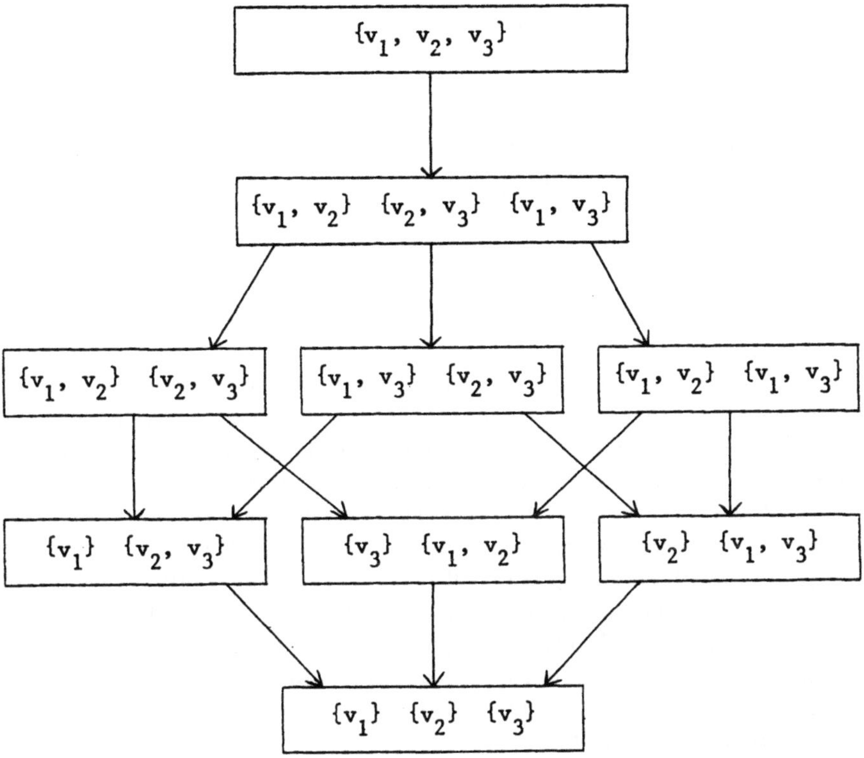

Figure 1.2. The Lattice of All Meaningful Decompositions or Models of the Overall System Defined in Table 1.1

any number of variables, as well as the lattice structure that the set of structure systems possesses (illustrated in Figure 1.2). This lattice structure is not extremely significant for this three-variable example. It is important, however, when dealing with more variables since the size of the set of structure systems grows rapidly and an intended "search" through this space is made dramatically easier by utilizing the lattice properties.

The "search" through the space of structure systems could be undertaken to determine which one of the structure systems adequately represents the overall behavior system. As GSPS is defined, "adequate representation" would have to be determined in the context of the particular investigation. The scope is easily illustrated, however, by considering Table 1.2. Table 1.2*a* gives the observations (for the behavior system of Table 1.1) associated with the two subsystems (or pairs of variables) $\nu_1\nu_2$ and $\nu_2\nu_3$. Table 1.2*b* gives the data for the two subsystems $\nu_1\nu_2$ and $\nu_1\nu_3$. In this case it is easy to show—and the implemented

Table 1.2. Two Possible Decompositions (Structure Systems) of the Overall System of Table 1.1

a

v_1	v_2	*Prob.*	v_2	v_3	*Prob.*
0	0	0	0	0	0
0	1	.5	0	1	.1
1	0	.1	1	0	.2
1	1	.4	1	1	.7

b

v_1	v_2	*Prob.*	v_1	v_3	*Prob.*
0	0	0	0	0	.2
0	1	.5	0	1	.3
1	0	.1	1	0	0
1	1	.4	1	1	.5

tools that have been referred to here are used to easily determine—that the structure system of Table 1.2*b* uniquely identifies the generative system of Table 1.1; that of Table 1.2*a* does not (see also Cavallo, 1980). In this case the investigator using this framework would quickly determine that the information associated with the relation between *age* (v_2) and *performance* (v_3) is not necessary for a perfect explanation of his data, while all other relational information is. It is worth emphasizing that this can be determined without the investigator's needing to be familiar with the details necessary for development of the software package. It is also worth emphasizing that this does *not* mean that *age* and *performance* are independent of each other, but rather only that the divergence from independence of *age* and *performance* can be controlled through consideration of the variable *sex*.

After the level of structure systems, GSPS recognizes higher-level systems called *metasystems*. Metasystems are defined by allowing the structure system itself to change in time, space, or any of the supporting attributes that have been utilized. (In the example above such changes in structure may occur over different groups of people or departments.) Metasystems at second and higher orders may be defined by allowing changes according to higher-level procedures (see Uyttenhove, 1979).

As mentioned previously, the dynamic aspect of GSPS as a representation of, and as an aid and mechanism for, modeling activity stems from the definition of *systems problems* based on the epistemological levels described. For the definitions of particular problems, the taxonomy of systems implicit in the hierarchy is extended somewhat to include other epistemological and methodological criteria, but these extensions are not especially relevant here (see Cavallo and Klir, 1978). Through use of these criteria, a set of well-defined system *types* is defined so that the set of all *particular* systems is partitioned and classified according to these types. Problems are basically of two kinds and as a class represent a characterization of different activities that are involved in systems investigations. The two problem kinds are: (1) Given a particular initial system, determine a particular terminal system of a specified type such that specified requirements (objectives and constraints) are satisfied (as in the illustration of structure system determination given above); (2) Given two particular systems, determine some property, specified by a set of requirements, of the relationship between the two systems (again referring to the example above, the two structure systems of Table 1.2 could be the given systems, and the requirements might specify determining which of the two is least ambiguous regarding the frequencies of the three variable states).

Each particular problem as specified within GSPS represents some aspect of the modeling process. For example, a particular *data system* may be given as an *initial* system, and a system type involving the level of *structure system* may be specified as *terminal* system type. These problems represent many classes of interpreted problems, such as theory construction in the social sciences, the programming problem in computer science, the problem of design in engineering, of simulation in computer modeling, or of "parsimonious description" as represented, say, in the use of factor analysis and certain transform methods. Problems of the first kind need not necessarily specify a terminal system type that is different from that of the initial system. For example, the problem of determining a suitable simplification of a system at any level (such as collapsing variables with minimal information loss) would be in this category.

Problems of the second kind encompass such situations as model verification, complexity comparisons with respect to specified criteria, and so on. One of the necessary steps in the elaboration of GSPS is the categorization of large problem classes, along with the assignment or development of methodological procedures suitable for their solution. Within the context of GSPS, each of these procedures represents an extension of information-processing capabilities and thus the material that specific system expertise will have available to work with. It is also important to observe that the association of methodological tools with problems is done in a manner in which the *problem* retains the main focus of attention and in which any given procedure (or set of procedures) plays a secondary role. From this point of view, a given problem—when dealt with by different procedures—may not have *a* solution. Rather, particular procedural results might represent a set of complementary results in light of which the particular problem (and its interpreted counterpart) may be considered. Reference to a number of works that deal with particular problems (and general techniques devoted to their solution), such as behavior determination, structure identification, and so on, may be found in Cavallo (1979*a*).

Chapter 5 illustrates the manner in which the structured methodological language that GSPS represents operates in conjunction with an overall investigation in a specific research area. The major point of emphasis is on pragmatist, and thus process, orientation and may be represented as in Figure 1.3. Just as

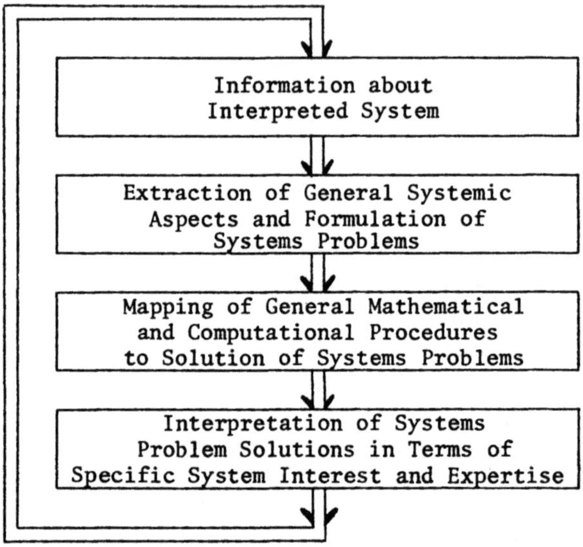

Figure 1.3. Schema Illustrating the Use of GSPS in Conjunction with a Specific Area of Research

important as problems that get "solved" is the well-specified approach toward investigation that imposes minimal restrictions and assumptions on the investigator. The use of the framework illustrates the continuous process of systems problem definition, solution, and interpretation leading to new problem definitions, and so on. The material in Chapter 5 demonstrates the process, with part of an ongoing utilization of GSPS, that augments the general system formulations with context from research dealing with conflict within nations.

REFERENCES

Ackoff, R. 1972. "Science in the Systems Age: Beyond I.E., O.R., and M.S." Paper delivered at joint AIEE, TIMS, ORSA meeting.

Atkin, R. 1974. *Mathematical Structure in Human Affairs.* New York: Crane-Russak.

Cavallo, R. 1979*a*. *The Role of Systems Methodology in Social Science Research.* Boston: Martinus Nijhoff.

——, ed. 1979*b*. *Systems Research Movement: Characteristics, Accomplishments, and Current Developments.* Louisville, Ky.: Society for General Systems Research.

——. 1980. "Reconstructability and Identifiability in the Evaluation of Structure

Hypotheses: An Issue in the Logic of Modeling." In B. Banathy, ed. *Systems Science and Science.* Louisville, Ky.: Society for General Systems Research.

Cavallo, R.E., and G.J. Klir. 1978. "A Conceptual Foundation for Systems Problem Solving." *International Journal of Systems Science* 9(2):219–36.

——. 1979*a*. "The Structure of Reconstructable Relations: A Comprehensive Study." In F. Pichler and R. Trappl, eds. *Progress in Cybernetics and Systems Research,* vol. 6. Washington, D.C.: Hemisphere. (Also in *Journal of Cybernetics* 9:399–413.)

——. 1979*b*. "Reconstructability Analysis of Multi-Dimensional Relations: A Theoretical Basis for Computer-Aided Determination of Acceptable Systems Models." *International Journal of General Systems* 5(2):143–71.

Cavallo, R.E., and F. Pichler. 1979. "General Systems Methodology: Designs for Intuition Amplification." In R. Ericson, ed. *Improving the Human Condition: Quality and Stability in Social Systems.* Basel: Birkhauser-Verlag.

Cortés, F., A. Przeworski, and J. Sprague. 1974. *Systems Analysis for Social Scientists.* New York: Wiley.

Gaines, B.R. 1978. "Progress in General Systems Research." In G. Klir, ed. *Applied General Systems Research: Recent Developments and Trends.* New York: Plenum Press.

Hayek, F.A. 1955. *The Counter Revolution of Science: Studies on the Abuse of Reason.* Glencoe, Ill.: Free Press.

Heisenberg, W. 1958. *Physics and Philosophy.* New York: Harper & Row.

Klir, G.J. 1969. *An Approach to General Systems Theory.* New York: Van Nostrand.

Lazarsfeld, P., ed. 1972. *Qualitative Analysis.* Boston: Allyn & Bacon.

Löfgren, L. 1978. "The Complexity Race." In G. Klir, ed. *Applied General Systems Research.* New York: Plenum Press.

Mesarovic, M.D., and Y. Takahara. 1975. *General Systems Theory: Mathematical Foundations.* New York: Academic Press.

Pask, G. 1969. "The Meaning of Cybernetics in the Behavioral Sciences." In J. Rose, ed. *Progress in Cybernetics.* New York: Gordon & Breach.

Piaget, J. 1970. *Genetic Epistemology.* New York: Columbia University Press.

Popper, K.R. 1972. *Objective Knowledge.* Oxford: Clarendon Press.

Przeworski, A., and H. Teune. 1970. *The Logic of Comparative Social Inquiry.* New York: Wiley.

Rescher, N. 1973*a*. *Conceptual Idealism.* Oxford: Basil Blackwell.

——. 1973*b*. *The Primacy of Practice.* Oxford: Basil Blackwell.

——. 1977. *Methodological Pragmatism.* New York: New York University Press.

Singer, J.D. 1971. *A General Systems Taxonomy for Political Science.* New York: General Learning Press.

Uyttenhove, H. 1979. "Computer-Aided Systems Modeling: An Assemblage of Methodological Tools for Systems Problem Solving." Ph.D. dissertation, State University of New York at Binghamton.

Wymore, A.W. 1976. *Systems Engineering Methodology for Interdisciplinary Teams.* New York: Wiley.

Zeigler, B.P. 1976. *Theory of Modeling and Simulation.* New York: Wiley.

2 ON THE USE OF STRUCTURED METHODOLOGIES IN GENERAL SYSTEMS RESEARCH

Franz Pichler

This chapter deals with the ways the influences of modern mathematics can be most profitably brought to bear on social systems modeling. The result of the discussion is that the best way is through the development of and focus of attention on "structured methodologies"; general systems research provides a mechanism to accomplish this development.

In ancient times mathematics was for the Greeks mainly an intellectual game that was not oriented toward applications. In later times this viewpoint changed considerably, primarily as a result of the consideration of mathematics as a tool for the development of scientific models. The fields of applications were mainly in physics and astronomy, and mathematics itself was strongly influenced by these applications. The subjects of real and complex analysis, of differential equations, and of functional analysis represent major developments of this influence. During the twentieth century mathematics has grown immensely. Not only have the topics of analysis and related fields such as topology been extensively developed, but also the algebraic fields have gained a great deal of attention and development. One of the "key discoveries" has been to emphasize the structural viewpoint of Bourbaki: to investigate a mathematical object (defined by axioms) in all its underlying structures and to relate it to similar objects. The result has been an interweaving of algebraic and analytical methods such that the sharp distinction between algebra and analysis has disappeared.

This development has taken place along with an extensive growth of engineering disciplines. As these disciplines have attempted to extend the scope and power of their investigations, mathematics has been increasingly applied to the design and the investigation of "machines." These new areas of applications have resulted in the development of different engineering-oriented mathematical "systems" theories. Examples are the theory of electrical networks, coding and information theory, and the theory of linear systems.

The social sciences, in which humans and the related environment are of central interest, are one of the latest products in the evolution of scientific disciplines. As in any discipline, social science research needs models that are reproducible and communicable in order to be used as objects of investigation to increase knowledge. Natural languages appear to be the natural first choice in which to formulate such models. Heisenberg's argument, referred to in Chapter 1, appears reasonable: natural languages, as designed for human uses, should also be among the best tools to help human beings in solving problems in the real world.

However, it is also logical that the different forms of our society today, with their high degree of what may be considered artificial or constructed aspects, are so complex and coupled together that natural languages alone no longer represent sufficient tools to develop, analyze, and synthesize models of social systems. Therefore, the natural languages must be enriched by such additional tools as mathematics and general systems theory, both closely related to the complexities introduced in modern societies.

These arguments are, in general, well accepted, and most social systems model-builders agree that the degree of "mathematization" in social systems modeling is at a somewhat preliminary level and has not been astoundingly successful. Most also agree that significant benefits from "mathematization" are still to be expected. It is a goal of this discussion to search for the reasons why "mathematization" in social systems modeling has lagged and to suggest improvements. We will come to the conclusion that the reason mathematics is not as successful as it could be is that "unstructured methodologies" have been favored by model-builders. For improvement we suggest the use of the "structured methodologies" that have been developed in the field of general systems theory; they are well represented in this book and are ready to be implemented in social science research (see also Wymore, 1976; Cavallo and Klir, 1978; Cavallo, 1979).

MATHEMATICS AND GENERAL SYSTEMS

As mentioned earlier, modern mathematics emphasizes the investigation of mathematical structure. Such emphasis is easy to identify by reading the titles of articles in recent mathematical journals. For example, some time ago papers with a title like "A remark to the theorem of X" were very common, while today favored titles look like "X as a generalization of Y." Although the investigation of structures in mathematics has been very successful from a mathematical point of view, the associated new ideas have not yet had the chance to be properly applied in the model-building area. Part of the reason may be that the subject of model building itself has not yet gained the attention in science that it deserves.

Only in recent years have we seen a change emerge; the activities of general systems research represent a main effort in this direction. General systems theory offers basic concepts that directly address the process of model building. The languages used in general systems research are a composition of natural languages along with mathematics and illustrations, so that interpretations in the real world may be more easily established. By investigating the kind of mathematics used in general systems theory, we discover that weakly structured mathematical objects such as sets, relations, directed graphs, networks, lattices, and automata are preferred. The "classical" mathematical objects, such as real numbers, linear vector-spaces, polynomials, real-valued functions, differential equations, and integrals, appear only in special cases. A similar result can be observed in problems. The "classical" types of problems, for example, "determine a 'nice' x such that $f(x) = y$," are in general systems theory often replaced by problems of the kind "to a given object X determine objects $X_1, X_2, \ldots, X_n$ and a composition $\circ$ between them, such that $X_1 \circ X_2 \circ \ldots \circ X_n$ simulates X." Such examples indicate the general systems theory emphasis on the use of modern mathematical concepts and the knowledge of different structures. This aspect of systems research has the powerful advantage that, if we follow the philosophy "construct for each general systems problem a mathematical problem that can serve as a paradigm," we should be able to lead modern mathematics to both new developments and new applications. Such an example is given in the theory of categories that has been recently related to general systems theory (Goguen, 1973; Arbib and Manes, 1977).

STRUCTURED METHODOLOGIES

We come now to the central part of this discussion. We want to investigate the kind of methodology that should be proposed to the social systems model-builder. First we discuss some preliminaries. We know from experience that, generally, significant efforts of a group of scientists are required to establish a new theoretical construct or method, and once the method is established and successfully in use, it takes a great deal of effort to convince the world that improvements are still possible (cf. Winch, 1958). These arguments may not apply as strongly to the natural sciences and the engineering-oriented disciplines where the demonstration of a new experiment or the design of a new "machine" are considered as achievements in themselves and may thus already be convincing. But it is reasonable to say that such arguments are applicable to the social sciences where demonstrating the building of new "machines" and performing experiments are comparatively difficult.

Man himself, since involved, can be considered a "resistor" in this process. We may demonstrate this by an example. As I have been told by colleagues whom I can trust, in economics certain mathematically oriented theories are still strongly in use and are the subjects of international discussion sessions (and have to be learned by students), even though they are demonstrably of very little help in solving real-world problems. But introducing new methods and convincing these people that the situation can be improved seems to be rather difficult. To improve this situation is to restate the theme of this chapter: *Avoid the use of "unstructured methodologies" and emphasize the application of what we will call "structured methodologies."*

An *unstructured methodology,* as we define it here, is a methodology that uses only mathematical objects having a given fixed mathematical structure. Examples are given by systems of linear equations and related methods for solution, systems of linear inequalities and related methods for solution and optimization, and linear regression methods or methods that deal with the construction of models by ordinary differential equations. In opposition to the concept of an unstructured methodology, we want to introduce the concept of a *structured methodology* as follows: a structured methodology consists of a collection of unstructured methodologies that are structurally related to each other. By that we know whether one methodology is a generalization or specialization of the other. We know whether one methodology is just a different representation of the others. And we also know in what ways use of one methodology can augment use of another. We could say that a structured methodology is nothing but a collection of methodologies that are organized under a common aspect of problem solving.

Let us try to point out some advantages that a structured methodology might provide. In applying a structured methodology, we can distinguish two steps: In step 1 we select a methodology that is the best one to deal with a perceived problem. This step enables us to adapt the model to a given problem. In step 2 we set the individual parameters of the selected method to achieve the best result. In the case of the application of an unstructured methodology, step 1 is replaced by a "random search" in the collection of unstructured methodologies that one has in mind; the adaptation to the problem is usually established by "fitting the parameters."

The conceptually simple idea presented here represents a development that is not yet in common use in social systems modeling. However, I believe that recognition of the importance of structured methodologies and adoption of the modeling approach will significantly improve the current situation's handling of complex, multiperspective problem situations.

Let us examine some further arguments that favor the use of structured methodologies. Step 1 (the selection of the appropriate method) can be done so that a given degree of accuracy in modeling the empirical data is guaranteed. Furthermore, this step enables us to select the best method to fit a given related theory (to get a model that "explains a lot"). Another argument is that a model that has been built by a structured methodology can be more easily adapted if necessary. In the case of an unstructured methodology, the adaptation can often be accomplished only by changing the parameters, while if a structured methodology is used, we can also achieve an adaptation by changing the methodology.

Where are the sources that offer structured methodologies? The answers to this question come from within the context of general systems research (see Klir, 1969, 1977; Cavallo, 1979; Wymore, 1976). But, as pointed out earlier, the underlying "philosophy" is also to be found in modern mathematics. Finally, we want to reiterate that improving the existing methods and making their importance transparent to social scientists is an important task. A few examples of successful applications already exist—for example, in the world-modeling efforts of Mesarovic and Pestel (1974).

SOCIAL SYSTEMS MODELING

A social system is defined here as any part of a reality in which human beings play an important role. This is obviously a rather wide concept of a social system. Clearly, in modeling, the factor "man" has to be carefully considered. In general he should play an important part. Unfortunately, many existing models of social systems do not consider this argument strongly enough. For example, look at the econometric models of economics. It is my opinion that many existing models fail on this point because they use unstructured methodologies. The knowledge of structured methodologies could bring improvements here since they offer the inclusion of the factor "man" in a sufficient way.

Let us illustrate this point with an example. Because of the limitations here, the example will be rather sketchy, and I can state the result in advance. By use of the structured methodology of dynamical system realization (Mesarovic and Takahara, 1975; Pichler, 1975), we are able to construct a dynamical model of a social system on a structural level that permits the factor "man" to be incorporated. The usual method would be to realize the dynamical model by a system of difference equations that imply given structures (e.g., the real numbers). These are often not appropriate (or not the most adequate) to describe the factor "man." For example, on a given social system we observe attributes out of two different classes with time as the running parameter. The one class of attributes describes the set of "setable" attributes, which we are able to control. The second class is the class of "observed" attributes. These attributes are thought to be controlled by the setable attributes. The problem is to find a "law" that describes the "causal" relation between the setable attributes during the period of the time of the experiment.

From a systems-theoretical view we are dealing here with an input-output relation, R, between the input-process, X, which is related to the setable attributes and the output-process, Y, which is related to the observed attributes: $R \subset X \times Y$. Each element of X is a time function x: $T \to A$; each element of Y is a time function y: $T \to B$. T means the assumed time set; A is the set of the different setable attributes (input states); B describes the set of the different observed attributes (output states). The problem of finding a "law" is systems-theoretically equivalent to the problem of finding a realization (a generative concept) of the relation R (the input-output process). To approach a solution we can use a structured methodology that is philosophically in line with the work of Mesarovic and Takahara (1975).

In the first step we develop individually for the sets A,B,X,Y,R, and T a hierarchy of admissible structured sets. Let us demonstrate this in short for T; T is in the most general sense just a set without any additional structure. Another possible structure for T would be that of an ordered set $(T,\leqslant)$. In addition, we could consider T to have the algebraic structure of an Abelian group $(T,+)$, which would give $(T,\leqslant,+)$ for the time set. Examples of rich-structured time sets would be $\mathbf{Z}$ (the integers) or $\mathbf{R}$ (the real numbers). We could, for example, agree that the whole hierarchy of time sets of our example would be considered as admissible.

In the second step we would search for a suited hierarchy of realization methods for R. First, the kind of "machine" that should realize R has to be determined. We assume we are satisfied if we are able to realize R by a dynamical system (ψ,β) with input and output. Then, depending on the structures chosen for A,B,X,Y,R, and T, respectively, different realization methods exist. From them we select those that we determine to be "best." The quality "best" might here be determined by the fact that the complexity of the method is within certain boundaries and that, on the other side, the method leads to a dynamical system (ψ,β) that is well suited to explain the "law" to be found (giving a close enough realization with respect to some criterion).

In the case of an unstructured method, the "random search" might, for example, lead to a system of difference equations. Then the structure of A,B,X,Y,R, and T would be already fixed ($A = \mathbf{R}^m$; $B = \mathbf{R}^p$; X, Y Hilbert spaces; R a linear relation; and $T = \mathbf{Z}$). The adaptation that takes account of the empirical situation could then be achieved only by selecting the proper parameters for the realization. Furthermore, in this case the "law" that is expressed by the systems of difference equations probably would be of little help in explaining the real system.

CONCLUSION

The main point emphasized here has been simple: the idea of a "structured methodology" is central for general systems methodologies, and it is an idea that is in line with the structural viewpoints of modern mathematics. Many will agree that the social systems modelbuilder does not yet fully appreciate and use such an approach, perhaps because the knowledge of such methods, where they do exist, is not yet widespread. The goal of this discussion has been to help the subject of structured methodologies receive in the future more of the attention that it deserves.

REFERENCES

Arbib, M.A., and E.G. Manes. 1977. "A Category-Theoretic Approach to Systems in A Fuzzy World." In W.E. Hartnett, ed. *Systems: Approaches, Theories, Applications.* Dordrecht, The Netherlands: Reidel.

Cavallo, R.E. 1979. *The Role of Systems Methodology in Social Science Research.* Boston: Martinus Nijhoff.

Cavallo, R.E., and G.J. Klir. 1979. "A Conceptual Foundation for Systems Problem Solving." *International Journal of Systems Science* 9 (2):219–36.

Goguen, J.A. 1973. "Categorical Foundations for General Systems Theory." In F. Pichler and R. Trappl, eds. *Advances in Cybernetics and Systems Research,* vol. 1. London: Transcripta Books.

Klir, G.J. 1969. *An Approach to General Systems Theory.* New York: Van Nostrand.

——. 1977. "Pattern Discovery in Activity Arrays." In W.E. Hartnett, ed. *Systems: Approaches, Theories, Applications.* Dordrecht, The Netherlands: Reidel.

Mesarovic, M., and E. Pestel, eds. 1974. *Multilevel Computer Model of World Development System.* Proceedings of the symposium held at IIASA 1974, vols. 1–6, Schloss Laxenburg, Austria.

Mesarovic, M., and Y. Takahara. 1975. *General Systems Theory: Mathematical Foundations.* New York: Academic Press.

Pichler, F. 1975. *Mathematische Systemtheorie: Dynamische Konstruktionen.* Berlin: Walter de Gruyter.

Winch, P. 1958. *The Idea of a Social Science.* New York: Humanities Press.

Wymore, A.W. 1976. *Systems Engineering Methodology for Interdisciplinary Teams.* New York: Wiley.

3 FUZZY SYSTEMS THEORY: *A Framework for the Analysis of Humanistic Systems*

Lotfi A. Zadeh

During the past decade the focus of research in systems theory has shifted increasingly toward the analysis of large-scale systems in which human judgment, perception, and emotions play an important role. Such *humanistic* systems are typified by socioeconomic systems, transportation systems, environmental control systems, food production systems, education systems, health care–delivery systems, criminal justice systems, information dissemination systems, and the like. The growing involvement of systems theorists in the analysis of large-scale systems—a trend that became discernible in the early sixties (Zadeh, 1962)—is a logical consequence of two concurrent developments: (1) the increasing interdependence of all sectors of modern industrial society and (2) the advent of powerful computers that make it possible to quantify, simulate, and analyze the behavior of systems involving hundreds and even thousands of interrelated variables.

Despite the widespread use of systems-theoretic methods in the modeling of large-scale systems, serious questions have been raised regarding the applicability of systems theory to the analysis of humanistic and, especially, socioeconomic

Research for this paper was supported by the National Science Foundation, Grants ENG76–84522 and MC76–06693.

systems. Hoos (1962), Berlinski (1976), and others have argued that systems analysis techniques are frequently misapplied, and that the armamentarium of system theorists has little, if anything, that is of value in attacking the extremely complex problems that arise in the analysis of large-scale systems that are imbued with uncertainty and imprecision.

As is frequently the case, the debate between systems theorists and their critics is not susceptible of definitive resolution. Personally, I tend to sympathize with the critics when they attack the mystique of systems analysis and deflate its exaggerated claims. On the other hand, I believe that, intrinsically, systems theory is a discipline of great potential importance and that, in the years to come, it will develop an effective body of concepts and techniques not only for the analysis of mechanistic systems—as it has done in the past—but, more important, for the analysis of humanistic systems in which the relations between system variables are too complex for description by differential, integral, or differential equations.

The systems theory of the future—the systems theory that will be applicable to the analysis of humanistic systems—is certain to be quite different in spirit as well as in substance from systems theory as we know it today. I will take the liberty of referring to it as *fuzzy systems theory* because I believe that its distinguishing characteristic will be a conceptual framework for dealing with a key aspect of humanistic systems—namely, the pervasive fuzziness of almost all phenomena that are associated with their external as well as internal behavior.

A rudimentary sketch of a conceptual framework for fuzzy systems theory has been presented in Zadeh (1973). A semantic point that is in need of clarification at this juncture is that fuzzy systems theory is not merely a theory of fuzzy systems, as described in Zadeh (1971), Negoita and Ralescu (1975), and other papers in the literature. Rather, it is a theory that allows an assertion about a system to be a fuzzy proposition (e.g., "System S is slightly nonlinear"). In this sense, then, fuzzy systems theory is not so much a precise theory of fuzzy systems as it is a fuzzy theory of both fuzzy and nonfuzzy systems. What this implies is that an assertion in fuzzy systems theory is normally not a theorem but a proposition that has a high degree of truth and, in addition, is informative in relation to a stated question.

At this juncture, fuzzy systems theory is not as yet an existing theory whose exposition can be found in the literature. What we have at present are merely the foundations for such a theory and a rough sketch of its structure. In what follows I shall attempt to convey my perception of some of the basic ideas that underlie fuzzy systems theory—ideas that, on further development, may eventually lead to a unified body of concepts and techniques for the analysis of humanistic systems.

BASIC CONCEPTS

Two related concepts are likely to play important roles in fuzzy systems theory: (1) the concept of a linguistic variable (Zadeh, 1973, 1975*a*) and (2) the concept of a fuzzy number (Zadeh, 1975*a*; Mizumoto and Tanaka, 1976; Nahmias, 1976; Jain, 1976*a*; Dubois and Prade, 1978). As its name implies, a *linguistic variable* is a variable whose values are expressed not as numbers but as words or sentences in a natural or synthetic language. For example, when the height of a person is characterized as not very tall, *not very tall* may be viewed as a linguistic value of the linguistic variable *Height.* Similarly, *quite high* is a linguistic value of the linguistic variable *Intelligence,* and *quite attractive* is a linguistic value of the linguistic variable *Appearance.*

Clearly, the use of words in place of numbers implies a low degree of precision in the characterization of the value of a variable. In some instances we elect to be imprecise because we do not need a higher degree of precision. In most cases, however, the imprecision is forced on us by the fact that there are no units of measurement for the attributes of an object and no quantitative criteria for representing the values of such attributes as points on an anchored scale.

Viewed in this perspective, the concept of a linguistic variable may be regarded as a device for systematizing the use of words or sentences in a natural or synthetic language for the purpose of characterizing the values of variables and describing their interrelations. In this role the concept of a linguistic variable serves a basic function in fuzzy systems theory, both in the representation of values of variables and in the characterization of truth-values of fuzzy propositions.

For the purpose of a brief exposition of the concept of a linguistic variable, we can conveniently focus our attention on a particular linguistic variable, say, Age, which may be viewed both as a numerical variable ranging over, say, the interval [0,150], and as a linguistic variable that can take the values *young, not young, very young, not very young, quite young, old, not very young and not very old,* and so forth. Each of these values may be interpreted as a label of a fuzzy subset of the universe of discourse $U = [0,150]$, whose base variable, u, is the generic numerical value of Age.

Typically, the values of a linguistic variable such as Age are built up of one or more *primary terms,* which are the labels of *primary fuzzy sets,*[1] together with a collection of modifiers and connectives that allow a composite linguistic value to be generated from the primary terms. Usually, the number of such terms is two, with one being an antonym of the other. For example, in the case of Age, the primary terms are *young* and *old,* with *old* being the antonym of *young.*

A basic assumption underlying the concept of a linguistic variable is that the meaning of the primary terms is context dependent, while the meaning of the modifiers and connectives is not. Furthermore, once the meaning of the primary terms is specified (or "calibrated") in a given context, the meaning of composite terms such as *not very young, not very young and not very old,* and so on, may be computed by the application of a semantic rule.

Typically, the *term set*—that is, the set of linguistic values of a linguistic variable—comprises the values generated from each of the primary terms, together with the values generated from various combinations of the primary terms. For example, in the case of Age, the following is a partial list of the linguistic values of Age:

young	old	not young nor old
not young	not old	not very young and not very old
very young	very old	young or old
not very young	not very old	not young or not old
quite young	quite old	etc.
more or less young	more or less old	
extremely young	extremely old	
etc.	etc.	

What is important to observe is that most linguistic variables have the same basic structure as Age. For example, on replacing *young* with *tall* and *old* with *short,* we obtain the list of linguistic values of the linguistic variable Height. The same applies to the linguistic variables Weight (*heavy* and *light*), Appearance (*beautiful* and *ugly*), Speed (*fast* and *slow*), Truth (*true* and *false*), and so forth, with the words in parentheses representing the primary terms.

As is shown in Zadeh (1975*a*), a linguistic variable may be characterized by an attributed grammar (Knuth, 1968; Lewis, Rosenkrantz, and Stearns, 1976) that generates the term set of the variable and provides a simple procedure for computing the meaning of a composite linguistic value in terms of the primary fuzzy sets that appear in its constituents. For example, if the meaning of *young* is characterized by a membership function such as

$$\mu_{\text{young}}(u) = \left[1 + \left(\frac{u}{30}\right)^2\right]^{-1}, \quad u \in [0, 150],$$

then the meaning of the linguistic value *not very young* can readily be computed to be

$$\begin{aligned}\mu_{\text{not very young}} &= 1 - \mu^2_{\text{young}}(u)\\ &= 1 - \left[1 + \left(\frac{u}{30}\right)^2\right]^{-2}.\end{aligned}$$

Similarly, the meaning of any linguistic value of *Age* may be expressed in terms of the membership functions of the primary fuzzy sets *young* and *old.*

The concept of a linguistic variable plays a particularly important role in fuzzy logic (Zadeh, 1975*b*; Bellman and Zadeh, 1977), where it serves to give an approximate characterization of the truth-value of a proposition (e.g., as *true, not true, very true, not very true,* and so on) or as the value of a quantifier (e.g., *many, most, few, almost all,* and so on). We shall have more to say about this application at a later point.

A concept that is closely related to that of a linguistic variable is the concept of a *fuzzy number* (Zadeh, 1975*a*), which is usually taken to be a convex fuzzy subset of the real line, R'. Fuzzy arithmetic—which may be viewed as a generalization of interval arithmetic (Moore, 1966)—is beginning to emerge as an important area of research within the theory of fuzzy sets (Mizumoto and Tanaka, 1976; Nahmias, 1976; Jain, 1976*a*; Dubois and Prade, 1978) and is likely to become a basic tool in the analysis of fuzzy systems in the years ahead.

Briefly, if x and y are fuzzy numbers and $*$ is a binary operation such as addition (+), subtraction (-), multiplication (×), division (/), max (∨), min (∧), and so forth, then by the application of the extension principle (Zadeh, 1975*a*), it can readily be shown that the fuzzy number z defined by

$$z = x * y$$

is given (as a fuzzy set) by

$$\mu_z(w) = V_{u,v}\mu_x(u) \wedge \mu_y(v), \qquad u, v, w \in R',$$

subject to the constraint

$$w = u * v,$$

where μ_x, μ_y, and μ_z denote the membership functions of x, y, and z, respectively, and $V_{u,v}$ denotes the supremum over u,v.

A basic property of fuzzy numbers, which has the effect of greatly simplifying operations on them, is that of *shape invariance.* This property was discovered by Nahmias (1976, 1978) and Misumoto and Tanaka (1976) and generalized by Dubois and Prade (1978).

As a simple illustration, if both x and y are *triangular* fuzzy numbers (i.e., μ_x and μ_y have the form of triangles) and $z = x + y$, then z will likewise be a triangular fuzzy number. Thus, if a triangular fuzzy number x is characterized by a triple $x = (x_1, x_2, x_3)$, in which x_1, x_2, and x_3 are the abscissae of the three vertices of the triangle [i.e., $\mu(x_1) = \mu(x_3) = 0, \mu(x_2) = 1$], then

$$x + y = (x_1, x_2, x_3) + (y_1, y_2, y_3)$$

is given by

$$x + y = (x_1 + y_1, x_2 + y_2, x_3 + y_3).$$

Similarly, if x and y are π-numbers (Zadeh, 1975*a*) (i.e., have piecewise-quadratic membership functions) characterized as ordered pairs $x = (\bar{x}, \beta_x)$, $y = (\bar{y}, \beta_y)$, where $\bar{x}$ and $\bar{y}$ are the abscissae of the peaks of x and y, and β_x and β_y are their respective bandwidths (i.e., the separations between the crossover points at which $\mu = 0.5$), then their sum, $x + y$, is a π-number characterized by

$$x + y = (\bar{x} + \bar{y}, \beta_x + \beta_y),$$

while their product, $x \times y$, is a fuzzy number characterized by

$$x \times y = (\bar{x} \times \bar{y}, \sigma_x + \sigma_y + \sigma_x \sigma_y),$$

where $\sigma_x = \beta_x / \bar{x}$ and $\sigma_y = \beta_y / \bar{y}$ denote the normalized bandwidths. Furthermore, for small σ_x and σ_y, $x \times y$ is, approximately, a π-number characterized by

$$x \times y \cong (\bar{x} \times \bar{y}, \sigma_x + \sigma_y).$$

The property of shape invariance, which holds exactly for addition and subtraction and approximately for other operations, makes it possible to represent fuzzy numbers in parametric form and thus translate various operations on fuzzy numbers into corresponding arithmetic operations on their parameters. This idea is developed at length in the recent work of Dubois and Prade (1978) and has the potential for many significant applications to the approximate analysis of both fuzzy and nonfuzzy systems.

Of particular note among the applications of linguistic variables and fuzzy numbers that have already been reported in the literature is the work of Wenstop (1975, 1976, 1977) on organization theory. Also of note is the rapidly expanding work on so-called fuzzy logic controllers (Mamdani, 1976; Mamdani and Assilian, 1975; Kickert and van Nauta Lemke, 1976; Jain, 1976*b*; Kickert and Mamdani, 1978; King and Mamdani, 1977; Rutherford and Carter, 1976; Ostergaard, 1977; Tong, 1976, 1977). What is surprising about the latter work is that fuzzy logic controllers have found considerable success not in the realm of humanistic systems—which motivated the conception of a linguistic variable and fuzzy logic—but in the very practical area of industrial process control.

Basic to the applications of linguistic variables and fuzzy numbers is the concept of a ***linguistic fuzzy-relational*** representation of dependencies in both fuzzy and nonfuzzy systems. Such representations are closely related to the concept of a branching questionnaire (Zadeh, 1976*b*) and subsume the rule-based systems (Winston, 1977) that are widely employed in artificial intelligence and related fields.

Typically, if $X_1, \ldots, X_n$ and Y are variables taking values in $U_1, \ldots, U_n$, and V, respectively, and R is a fuzzy relation from $U_1, \ldots, U_n$ to V, then a linguistic fuzzy-relational representation of R has the form of a tableau such as shown in Table 3.1, in which the r_{ij} and r_i are labels of fuzzy subsets of $U_1, \ldots, U_n, V$. In general, the r_{ij} and r_i are fuzzy numbers or the linguistic values of attributes, grades of membership, truth-values, probability-values, or possibility-values. Thus, typically, a row of R may contain entries exemplified by the following: (150,5) (a fuzzy number with peak value 150 and bandwidth 5 representing an approximate value of an attribute, say, Weight); *not very young* (a linguistic value of Age); *very low* (a linguistic value of the grade of membership in a fuzzy set, e.g., the class of creative people); *quite true* (a linguistic value of the fuzzy predicate Healthy); *not very likely* (a linguistic value of the probability of a fuzzy or nonfuzzy event, e.g., inheriting a fortune); (0.4, 0.1) (a fuzzy number representing the probability of, say, living beyond the age of 60); and *quite possible* (a linguistic value of the possibility of, say, carrying seven passengers in a five-passenger car).

As shown in Zadeh (1977*a, b*), a tableau of the form shown in Table 3.1 *translates* into an $(n + 1)$-ary fuzzy relation expressed by

$$\begin{aligned} \tilde{R} = {} & r_{11} \times r_{12} \times \cdots \times r_{1n} \times r_1 + \\ & r_{21} \times r_{22} \times \cdots \times r_{2n} \times r_2 + \cdots + \\ & r_{m1} \times r_{m2} \times \cdots \times r_{mn} \times r_m, \end{aligned} \tag{3.1}$$

where + denotes the union of fuzzy sets, which for fuzzy subsets A, B of U is defined by

$$\mu_{A+B}(u) = \mu_A(u) \vee \mu_B(u), \qquad u \in U,$$

while $\times$ is the Cartesian product, defined by

$$\mu_{A \times B}(u, v) = \mu_A(u) \wedge \mu_B(v), \qquad u \in U, \qquad v \in V.$$

Thus, through the use of (3.1), a relation R, whose elements are fuzzy numbers

Table 3.1. Tableau of R

X_1	X_2		X_n	Y
r_{11}	r_{12}	$\cdots$	r_{1n}	r_1
r_{21}	r_{22}	$\cdots$	r_{2n}	r_2
$\vdots$	$\vdots$		$\vdots$	$\vdots$
r_{m1}	r_{m2}	$\cdots$	r_{mn}	r_m

and/or the values of linguistic variables, may be translated into a fuzzy relation $\tilde{R}$, whose elements are $(n + 1)$-tuples in $U_1 \times U_2 \times \cdots \times U_n \times V$.

We speak of the *translation* of R into $\tilde{R}$ because $\tilde{R}$ defines the meaning of R as a fuzzy $(n + 1)$-ary relation in $U_1 \times U_2 \times \cdots \times U_n \times V$. Having $\tilde{R}$, we can define the effect of various operations that may be performed on R. In particular, the knowledge of $\tilde{R}$ provides a basis for an interpolation of R for values of $X_1, \ldots, X_n$ that are not tabulated in the tableau of R. More specifically, assume that we wish to find the value of Y corresponding to the linguistic values $X_1 = s_1$, $X_2 = s_2, \ldots, X_n = s_n$, where the n-tuple $(s_1, \ldots, s_n)$ does not appear as a row in the tableau of R. To this end, compute the Cartesian product

$$S = s_1 \times \cdots \times s_n$$

as a fuzzy relation in $U_1 \times \cdots \times U_n$. Then, as shown in Zadeh (1977*a*), the corresponding value of Y is given approximately by the formula

$$Y = \tilde{R} \circ S, \tag{3.2}$$

where $\circ$ denotes the operation of composition.[2] If necessary, the value of Y yielded by (3.2) may be approximated by a linguistic value of Y through the use of linguistic approximation (Zadeh, 1975*a*; Wenstop, 1976; Procyk, 1976).

Linguistic fuzzy-relational representations are particularly relevant to the definition of complex concepts (Zadeh, 1976*a*), characterization of control strategies (Mamdani and Assilian, 1978; Tong, 1977), pattern classification (Zadeh, 1977*a*), and diagnostics (Sanchez, 1977). Furthermore, they provide a basis for the characterization of dependencies in which the sources of imprecision comprise both fuzzy and random variables. In such cases the tableau of R may be expressed in the canonical form shown in Tables 3.2*a* and 3.2*b*, in which the r_i are pointers to relations of the form shown in Table 3.2*b*, with $F_1, \ldots, F_k$ representing specified fuzzy sets and p_r, $r = 1, \ldots, \ell$, denoting a linguistic probability that an object characterized by the rth row of R has the grade of membership μ_{r1} in F_1, μ_{r2} in $F_2, \ldots, \mu_{rk}$ in F_k, with $\mu_{r1}, \ldots, \mu_{rk}$ expressed as fuzzy subsets of the unit interval.

As an elementary illustration,[3] assume that $X_1, \ldots, X_n$ represent a set of symptoms (e.g., $X_1 \triangleq$ body temperature, $X_2 \triangleq$ pulse rate, $X_3 \triangleq$ presence of pain); $F_1, \ldots, F_k$ are possible illnesses corresponding to the symptoms in question (e.g., flu and bronchitis); and p_r is the probability that a person has the illnesses $F_1, \ldots, F_k$ to the degrees $\mu_{r1}, \ldots, \mu_{rk}$ simultaneously, given the linguistic values of the symptoms $X_1, \ldots, X_n$. In this case, Tables 3.2*a* and 3.2*b* may have the form shown in Tables 3.3*a, b,* and *c.* Thus, if the temperature is *high* (with *high* regarded as a primary term of the linguistic variable Temperature), the pulse rate is given approximately by the fuzzy number (80,5), and the truth-value of the fuzzy predicate Pain is *not true,* where *true* is a primary term

Table 3.2. Tableau of R and the Pointer r_i

a

R	X_1	X_2	$\cdots$	X_n	Y_1
	r_{11}	r_{12}	$\cdots$	r_{1n}	r_1
	r_{21}	r_{22}	$\cdots$	r_{2n}	r_2
	$\vdots$	$\vdots$		$\vdots$	$\vdots$
	r_{m1}	r_{m2}	$\cdots$	r_{mn}	r_m

b

r_i	F_1	F_2	$\cdots$	F_k	P
	μ_{11}	μ_{12}	$\cdots$	μ_{1k}	p_1
	μ_{21}	μ_{22}	$\cdots$	μ_{2k}	p_2
	$\vdots$	$\vdots$		$\vdots$	$\vdots$
	$\mu_{\ell 1}$	$\mu_{\ell 2}$	$\cdots$	$\mu_{\ell k}$	p_ℓ

Note: $i = 1, \ldots, m$ (i as a subscript or superscript is implicit in all of the entries in the tableau of r_i).

of the linguistic variable Truth; then with linguistic probability ***very likely,*** the diagnosis is "flu to the degree *low* and bronchitis to the degree ***high***"; with linguistic probability ***unlikely,*** the diagnosis is "flu to the degree ***high*** and bronchitis to the degree *low*"; and, with probability ***unlikely,*** the diagnosis is "flu to the degree *low* and bronchitis to the degree *low*." In these expressions, *low* is a primary term for the linguistic variable Grade, ***high*** is the antonym of *low*; ***likely*** is a primary term for the linguistic variable Probability, and ***unlikely*** is the antonym of likely.[4]

In the foregoing discussion, we have restricted our attention to the representation of attribute dependencies in the case of a single object. In systems analysis, however, our main concern is with the representation of the collective behavior of a group of interacting objects rather than with the behavior of a single object. Thus, the question is, How can the fuzzy-relational representation of the constituents of a system be combined to form a fuzzy-relational representation of the system as a whole? Although some aspects of this problem have been studied and reported in the literature (Zadeh, 1973, 1971; Negoita and Ralescu, 1975; Kaufmann, 1975), it remains, at this juncture, a largely unexplored basic issue in fuzzy systems theory.

Table 3.3. Tableau of R, the Pointer r_1, and the Pointer r_2

a

Temp.	*Pulse*	*Pain*	*Y*
High	(80,5)	Not true	r_1
High	(70,5)	True	r_2
.	.	.	.
Normal	(60,5)	False	r_m

b

r_1	*Flu*	*Bronchitis*	*P*
	Low	High	Very likely
	High	Low	Unlikely
	.	.	.
	Low	Low	Unlikely

c

r_2	*Flu*	*Bronchitis*	*P*
	Low	High	Unlikely
	High	Low	Very likely
	.	.	.
	High	High	Unlikely

FUZZY SYSTEMS THEORY AND FUZZY LOGIC

An essential distinction between classical systems theory, on the one hand, and fuzzy systems theory, on the other, relates to a basic difference in their underlying logics. Thus, classical systems theory, like most other mathematical theories, rests on the foundation of two-valued logic, which implies that the propositions asserted in classical systems theory are either true—in which case they are theorems—or false.

In contrast, fuzzy systems theory is based on *fuzzy logic* (Zadeh, 1975*b*, 1977*c*; Bellman and Zadeh, 1977)—that is, on a logic with fuzzy truth-values and rules of inference that are approximate rather than exact. In this logic a fuzzy proposition is usually associated with a fuzzy truth-value (e.g., the truth-value of the proposition "X is much larger than Y" may be *not very true*). Thus, what matters in fuzzy logic are not theorems but propositions that, though fuzzy, are informative and have a high degree of truth. Such propositions are, in general, inferences drawn from fuzzy premises, and their degree of truth is determined in an approximate manner by the linguistic truth-values of the premises from which they are inferred.

A concept that plays a basic role in fuzzy logic is that of a *possibility distribution* (Zadeh, 1978).[5] Thus, if X is a variable taking values in a universe of discourse, U, then a possibility distribution, Π_X, associated with X, is a fuzzy subset F of U such that the *possibility* that X may take a value $u \in U$ is numerically equal to the grade of membership of u in F, that is,

$$\begin{aligned} \text{poss}\{X = u\} &= \pi_X(u), \qquad u \in U \\ &= \mu_F(u), \end{aligned} \tag{3.3}$$

where μ_F is the membership function of F and $\pi_X(u)$ is termed the *possibility distribution function* associated with Π_X.

Given the possibility distribution of X, the possibility of X taking a value in a specified subset A of U is given by

$$\text{poss}\{X \in A\} = V_u \pi_X(u), \qquad u \in U, \tag{3.4}$$

and, if A is a fuzzy subset characterized by its membership function μ_A, then

$$\text{poss}\{X \text{ is } A\} = V_u \pi_X(u) \wedge \mu_A(u), \qquad u \in U. \tag{3.5}$$

As an illustration, if *small* is a specified fuzzy subset of the real line, then the proposition "X is small" translates into what is called a *possibility assignment equation*

$$\Pi_X = \text{small},$$

which implies that

$$\text{poss}\{X = u\} = \mu_{\text{small}}(u), \qquad u \in (-\infty, \infty).$$

Furthermore, if we know that X is small, then, by (3.5), the possibility that X is not very small is given by

$$\text{poss}\{X \text{ is not very small}\} = V_u \{\pi_{\text{small}}(u) \wedge [1 - \mu^2_{\text{small}}(u)]\}. \tag{3.6}$$

More generally, a proposition of the form "X is F," where X is a variable taking values in U, and F is a fuzzy subset of U, translates into a possibility assignment equation of the form

$$\Pi_X = F; \tag{3.7}$$

which implies that

$$\text{poss}\{X = u\} = \mu_F(u), \qquad u \in U.$$

The concept of a possibility distribution provides a basis for the translation of propositions expressed in a natural language into expressions in a synthetic meaning representation language called PRUF (Zadeh, 1977*b*). Thus, an expression in PRUF is, in general, a procedure that acts on a collection of relations in a data base and returns a possibility distribution. For example, given the definition of *small* as a fuzzy subset of the real line, the translation of the proposition "X is not very small" may be expressed as ($\rightarrow$ stands for "translates into")

$$X \text{ is not very small} \rightarrow \Pi_X = 1 - (\text{small}^2),$$

which means that the proposition in question induces a possibility distribution of X that may be obtained from the fuzzy set small by squaring small (i.e., squaring its membership function) and forming its complement (i.e., subtracting the resulting membership function from unity). It should be noted that it is this translation of the proposition "X is not very small" that is employed in (3.6) to compute the possibility that X is not very small.

Another example of a basic translation rule in PRUF is the *rule of conjunctive composition,* which asserts that if

$$X \text{ is } F \rightarrow \Pi_X = F, \qquad F \subset U$$

and

$$Y \text{ is } G \rightarrow \Pi_Y = G, \qquad G \subset V,$$

then

$$X \text{ is } F \text{ and } Y \text{ is } G \rightarrow \Pi_{(X,Y)} = F \times G, \tag{3.8}$$

where $\Pi_{(X,Y)}$ is the joint possibility distribution of the variables X and Y, and $F \times G$ is the Cartesian product of the fuzzy sets F and G. Similarly, the *rule of conditional composition* states:

$$\text{If } X \text{ is } F, \text{ then } Y \text{ is } G \rightarrow \Pi_{Y(X)} = \bar{F}' \oplus \bar{G}, \tag{3.9}$$

where $\Pi_{Y(X)}$ is the conditional possibility of Y given X, and the fuzzy set $\bar{F}' \oplus \bar{G}$ is defined by Zadeh (1975*a*):

$$\mu_{\bar{F}' \oplus \bar{G}}(u, v) = 1 \wedge [1 - \mu_F(u) + \mu_G(v)], \qquad u \in U, \qquad v \in V. \tag{3.10}$$

When a proposition is qualified by a truth-value, as in "X is small is very true," its translation is governed by the *rule of truth qualification,* which asserts:

$$\text{If } X \text{ is } F \rightarrow \Pi_X = F,$$

then

$$X \text{ is } F \text{ is } \tau \rightarrow \Pi_X = F^+, \tag{3.11}$$

where τ is a linguistic truth-value, μ_F and μ_T are the membership functions of F and τ, respectively, and

$$\mu_{F^+}(u) = \mu_\tau[\mu_F(u)], \quad u \in U. \tag{3.12}$$

Thus, the association of a truth-value, τ, with a proposition of the form "X is F," has the effect of modifying the possibility distribution of X in the manner defined by (3.12).

Our brief discussion of some of the elementary translation rules in fuzzy logic is intended to illustrate the basic principles by which a fairly complex composite proposition describing a relation between two or more variables may be translated into an assertion concerning the joint or conditional possibility distribution of the variables in question. For example, if the relation between X, Y, and Z is described by the proposition

"If X is small and Y is large, then Z is very small," (3.13)

then, by the use of (3.8) and (3.9), the proposition in question may be translated into a possibility assignment equation that expresses the conditional possibility distribution $\Pi_{Z(X,Y)}$ as a function of the primary fuzzy sets *small* and *large*. Given this distribution, then, we can find the possibility distribution of Z from a specification of the possibility distributions of X and Y.

More generally, if Y is the output of a system, S, if $X_1, \ldots, X_n$ are the inputs to S, and if the dependence of Y on $X_1, \ldots, X_n$ is described by a collection of propositions of the form (3.13), then, by the application of translation rules of fuzzy logic, we can compute the possibility distribution of Y as a function of the possibility distributions of $X_1, \ldots, X_n$.

In the foregoing exposition we have limited our attention to those aspects of fuzzy logic that are of relevance to the analysis of fuzzy systems—that is, systems that are characterized by fuzzy input-output relations. What should be recognized, however, is that, counter to what one might expect, fuzzy logic also has an important role to play in the analysis of nonfuzzy systems. In this role it provides (1) a language (PRUF) for an approximate characterization of nonfuzzy input-output relations and (2) a method for assessing the consistency of such characterizations with the data resident in a data base.

As a very simple illustration of this application, suppose that S is a memoryless system with input X and output Y, and that we have a tabulation of Y as a function of X. Assume that the dependence of Y on X is expressed in a summarized form by the proposition

"If X is small then Y is large, and if X is not small then
Y is not very large." (3.14)

By using the translation rules (3.8) and (3.9), the proposition in question translates into the possibility assignment equation (Zadeh, 1978)

$$\Pi_{Y(X)} = [\overline{\text{small}'} \oplus \overline{\text{large}}] \cap [\overline{\text{small}} \oplus \overline{(\text{large}^2)'}], \tag{3.15}$$

which implies that the conditional possibility distribution function of Y given X is expressed by

$$\pi_{Y(X)}(u, v) = [1 - \mu_{\text{small}}(u) + \mu_{\text{large}}(v)] \wedge [\mu_{\text{small}}(u) + 1 - \mu^2_{\text{large}}(v)]. \tag{3.16}$$

Now let (x_i, y_i) be the entry in the ith row of the tabulation of Y as a function of X. Substituting this pair into (3.16), we obtain its consistency, γ_i, with the given proposition—that is,

$$\gamma_i = \pi_{Y(X)}(x_i, y_i). \tag{3.17}$$

Then, using (3.17), a conservative assessment of the degree of consistency between the given tabulation of Y as a function of X and its summary as expressed by (3.16) may be defined as

$$\gamma = \min_i \pi_{Y(X)}(x_i, y_i), \tag{3.18}$$

with i ranging over the rows of the tabulation. It is understood, of course, that other, less conservative, measures of consistency may be preferable in some applications.

In summary, fuzzy logic serves three essential functions in fuzzy systems theory. First, it provides a language (PRUF) for the characterization of input-output relations of both fuzzy and nonfuzzy systems. Second, it provides a system for drawing approximate conclusions from imprecise data. And finally, it provides a method for assessing the goodness of a model or, equivalently, a summary of the behavior of a system that is too complex or too ill-defined to be susceptible to analysis by conventional techniques.

CONCLUDING REMARKS

Fuzzy systems theory, as outlined in our exposition, represents a rather sharp break with the tradition of precision and exactitude in classical systems theory. Based as it is on fuzzy rather than two-valued logic, fuzzy systems theory does not aim at the discovery of precise assertions about the behavior of complex systems. Rather, it aims at an accommodation with the pervasive imprecision of real-world systems by abandoning the unattainable goals of classical systems theory and adopting instead a conceptual framework that is tolerant of imprecision and partial truths.

Fuzzy systems theory is not as yet an existing theory. What we have at present are merely parts of its foundation. Nevertheless, even at this very early stage of its development, fuzzy systems theory casts some light on the processes of approximate reasoning in human decision making, planning, and control. Furthermore, in the years ahead, it is likely to develop into an effective body of concepts and techniques for the analysis of large-scale humanistic as well as mechanistic systems.

NOTES

1. In the case of humanistic systems, primary fuzzy sets play a role that is somewhat analogous to that of physical units in the case of mechanistic systems.

2. Equation (3.2) implies that, in terms of the membership functions of $\tilde{R}$ and S, the membership function of Y is given by

$$\mu_Y(v) = V_{u_1, \ldots, u_n} [\mu_{\tilde{R}}(u_1, \ldots, u_n, v) \wedge \mu_S(u_1, \ldots, u_n)],$$

where $V_{u_1, \ldots, u_n}$ denotes the supremum over $u_1 \in U_1, \ldots, u_n \in U_n$.

3. Needless to say, this example does not pretend to be a realistic representation of the relation between the symptoms in question and the corresponding diseases.

4. If A is a fuzzy subset of the unit interval, as is true of both *low* and *likely*, then the membership function of the antonym of A is related to that of A by

$$\mu_{\text{ant}(A)}(v) = \mu_A(1 - v), \qquad v \in [0,1].$$

5. Intuitively, *possibility* relates to feasibility, ease of attainment, and compatibility, whereas *probability* relates to likelihood, frequency, and strength of belief. In contrast to the concept of probability, the concept of possibility is nonstatistical in nature.

REFERENCES

Bellman, R.E., and L.A. Zadeh, 1977. "Local and Fuzzy Logics." In J.M. Dunn and G. Epstein, eds. *Modern Uses of Multiple-Valued Logic.* Dordrecht, The Netherlands: Reidel.

Berlinski, D. 1976. *On Systems Analysis.* Cambridge, Mass.: MIT Press.

Dubois, D., and H. Prade. 1978. "Fuzzy Algebra, Analysis, Logics." Technical Report 78-13, Purdue University, School of Electrical Engineering.

Hoos, I. 1962. *Systems Analysis in Public Policy.* Berkeley: University of California Press.

Jain, R. 1976*a*. "Tolerance Analysis Using Fuzzy Sets." *International Journal of Systems Science* 7:1393-1401.

——. 1976*b*. "Outline of an Approach for the Analysis of Fuzzy Systems." *International Journal of Control* 23:627-40.

Kaufmann, A. 1975. *Introduction to the Theory of Fuzzy Subsets,* vol. 3.:

Applications to Classification and Pattern Recognition, Automata and Systems, and Choice of Criteria. Paris: Masson.

Kickert, W.J.M., and E.H. Mamdani. 1978. "Analysis of a Fuzzy Logic Controller." *Fuzzy Sets and Systems* 1:29–44.

Kickert, W.J.M., and H.R. van Nauta Lemke. 1976. "Application of a Fuzzy Controller in a Warm Water Plant." *Automatica* 12:301–08.

King, P.J., and E.H. Mamdani. 1975. "The Application of Fuzzy Control Systems to Industrial Processes." *Proceedings of the 6th IFAC World Congress,* Boston. (Also in *Automatica* 13(1977):235–42.)

Knuth, D.E. 1968. "Semantics of Context-Free Languages." *Mathematical Systems Theory* 2:127–45.

Lewis, P.M., D.J. Rosenkrantz, and R.E. Stearns. 1976. "Attributed Translations." *Journal of Computer and System Sciences* 9:279–307.

Mamdani, E.H. 1976. "Advances in the Linguistic Synthesis of Fuzzy Controllers." *International Journal of Man-Machine Studies* 8:669–78.

Mamdani, E.H., and S. Assilian. 1975. "An Experiment in Linguistic Synthesis with a Fuzzy Logic Controller." *International Journal of Man-Machine Studies* 7:1–13.

Mizumoto, M., and K. Tanaka. 1976. "Algebraic Properties of Fuzzy Numbers." *Proceedings of the International Conference on Cybernetics and Society,* Washington, D.C.

Moore, R.E. 1966. *Interval Analysis.* Englewood Cliffs, N.J.: Prentice-Hall.

Nahmias, S. 1976. "Fuzzy Variables." Technical Report 33, University of Pittsburgh, Department of Industrial Engineering, Systems Management Engineering and Operations Research.

——. 1978. "Fuzzy Variables." *Fuzzy Sets and Systems* 1:97–110.

Negoita, C.V., and D.A. Ralescu. 1975. *Applications of Fuzzy Sets to Systems Analysis.* Basel: Birkhauser Verlag.

Ostergaard, J.-J. 1977. "Fuzzy Logic Control of a Heat Exchanger Process." In M.M. Gupta, G.N. Saridis, and B.R. Gaines, eds. *Fuzzy Automata and Decision Processes.* New York: North-Holland.

Procyk, T.J. 1976. "Linguistic Representation of Fuzzy Variables." Queen Mary College, Fuzzy Logic Working Group, London.

Rutherford, D., and G.A. Carter. 1976. "A Heuristic Adaptive Controller for a Sinter Plant." *Proceedings of the 2nd IFAC Symposium on Automation in Mining, Mineral and Metal Processing,* Johannesburg.

Sanchez, E. 1977. "Solutions in Composite Fuzzy Relation Equations: Application to Medical Diagnosis in Brouwerian Logic." In M.M. Gupta, G.N. Saridis, and B.R. Gaines, eds. *Fuzzy Automata and Decision Processes.* New York: North-Holland.

Tong, R.M. 1976. "Analysis of Fuzzy Control Algorithms Using the Relation Matrix." *International Journal of Man-Machine Studies* 8:679–86.

——. 1977. "A Control Engineering Review of Fuzzy Systems." *Automatica* 13: 559–69.

Wenstøp, F. 1975. "Application of Linguistic Variables in the Analysis of Organizations." Ph.D. thesis, University of California, Berkeley, School of Business Administration.

——. 1976. "Deductive Verbal Models of Organizations." *International Journal of Man-Machine Studies* 8:293–311.

——. 1977. "Fuzzy Sets and Decision-Making." *California Engineer* 16:20–24.

Winston, P.H. 1977. *Artificial Intelligence.* Reading, Mass.: Addison-Wesley.

Zadeh, L.A. 1962. "From Circuit Theory to System Theory." *Proceedings of the IRE* 50:856–65.

——. 1971. "Toward a Theory of Fuzzy Systems." In R.E. Kalman and N. DeClaris, eds. *Aspects of Networks and Systems Theory.* New York: Holt, Rinehart & Winston.

——. 1973. "Outline of a New Approach to the Analysis of Complex Systems and Decision Processes." *IEEE Transactions on Systems, Man and Cybernetics,* SMC-3:28–44.

——. 1975*a*. "The Concept of a Linguistic Variable and Its Application to Approximate Reasoning." Part I, *Information Sciences* 8:199–249; Part II, *Information Sciences* 8:301–57.

——. 1975*b*. "Fuzzy Logic and Approximate Reasoning (In Memory of Grigore Moisil)." *Synthese* 30:407–28.

——. 1976*a*. "The Concept of a Linguistic Variable and Its Application to Approximate Reasoning." Part III, *Information Sciences* 9:43–80.

——. 1976*b*. "A Fuzzy-Algorithmic Approach to the Definition of Complex or Imprecise Concepts." *International Journal of Man-Machine Studies* 8:249–91.

——. 1977*a*. "Fuzzy Sets and Their Application to Pattern Classification and Clustering Analysis." In J. Van Ryzin, ed. *Classification and Clustering.* New York: Academic Press.

——. 1977*b*. "PRUF – A Meaning Representation Language of Natural Languages." Memorandum 77/61, University of California, Berkeley, Electronics Research Laboratory.

——. 1977*c*. "A Theory of Approximate Reasoning." Memorandum M77/58, University of California, Berkeley, Electronics Research Laboratory.

——. 1978. "Fuzzy Sets as a Basis for a Theory of Possibility." *Fuzzy Sets and Systems* 1:3–28.

II GENERAL METHODOLOGICAL APPROACHES:
The Theory/Data Interface in Social Systems Investigation

Chapter 1 presented issues that characterized systems methodology as mediating between context-oriented and abstract languages, or, in process terms, between description and interpretation. In Cavallo (1979), this characterization is elaborated in detail. A basic aspect of this characterization, illustrated in Figure II.1, is that, as mathematics and computer science have each come to be considered as independent fields of study, many results in these fields do not originate from context-oriented problems and, in general, are not immediately useful for their solution. As each discipline or field of study has also developed its own highly specialized language, "interdisciplinary" approaches to the meaningful merging of expertise from various pairings of these languages have not, in general, proven very successful. This appears to be especially true for language pairings representing extremes of abstraction and interpretation. Systems methodology as a level of abstraction that deals with problem characterization rather than specific content serves the role of defining the utilization of various abstract results and procedures for classes of problems meaningful over a broad spectrum of content, as well as that of motivating abstract language research with high potential utility, and of providing an effective nondisciplinary framework for communication.

Each of the chapters in Part II presents developments in this area, and each provides mechanisms that are described in a sufficiently abstract manner to

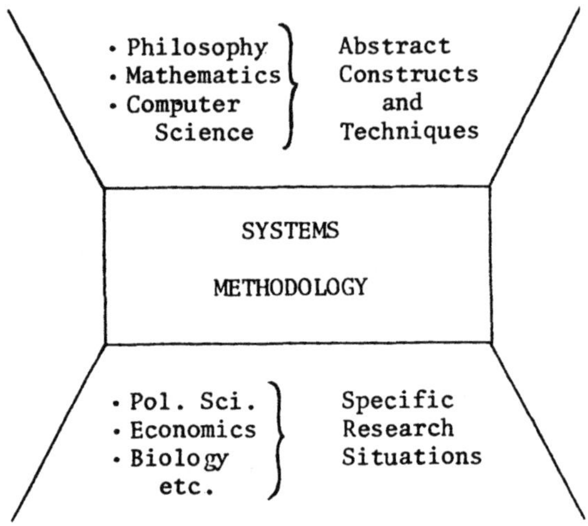

Figure II.1. Systems Methodology as Process Connecting Abstractions and Objects

allow utilization of general mathematical and data-processing results. At the same time each chapter is immediately concerned with systems problems, such as that of structure determination, that are easily seen to be relevant to the investigation of any system. It is also important to note that the methodological developments deal with large but integrated classes of systems problems and are also general enough to leave room for–and demand–interpretation of results in terms of specific system expertise.

In this regard the use of mathematical and computational tools in terms of systems concepts and problems represents a nonrestrictive extension of the investigative capabilities of researchers in specific areas. I feel that these "automated," but at the same time "not-automated," implemented characterizations of the modeling process represent an epistemological innovation that will have deep and long-lasting effects on our abilities to deal and effectively interact with complex humanistic systems.

4 THE METHODOLOGY OF *Q*-ANALYSIS APPLIED TO SOCIAL SYSTEMS

Ronald H. Atkin

Some of the methodological consequences of using the basic notion of set-membership to define "scientific" or "hard" data are discussed in this chapter. Chiefly because of the Russell theory of types (but not exclusively because of that), a need exists for a well-defined hierarchy of data sets. Such a hierarchy is defined in terms of cover sets (rather than by partitions) and is expressed by a set of mathematical relations between the finite data sets. The structure of the data is then identified by the simplicial complexes that represent these relations, and these will contain the static backcloth, $S(N)$, for that data. Some simple connectivity properties of a typical complex are listed, and an illustration of the methodological technique is provided by examples taken from an earlier study of an area in the town of Southend-on-Sea and from a current regional project. The importance of the structure (q-connectivities) for the dynamics of (generalized) traffic on the backcloth, $S(N)$, is then examined and critically compared with regression analysis.

HARD DATA AND SET-MEMBERSHIP

We begin by examining the consequences of the proposition that scientific data result from observing set-membership. This kind of data is referred to as "hard data," and (by default) all other observations are called "soft data."

The idea of set-membership is a nontrivial constraint on the acceptable methodology and is more fundamental than the earlier Kelvin view that "data is what is measurable." Here we are effectively saying that "measuring" is equivalent to the identification of set-membership. We are replacing Kelvin's idea by the following statement: The only acceptable data are hard data.

For the developments of this chapter, collecting hard data presupposes that sets are *well defined.* By this we mean that, given a set X, it is possible to say of some proposed element either that it is or that it is not a member of X. Such decisions are binary, being yes/no or 1/0 type. In the physical laboratory these binary decisions are commonly delegated to the physical instruments themselves. It is important to realize that these instruments, which take over many of the day-to-day problems of the scientist, are the result of many years of work and development devoted to quite specific problems. Thus, an ammeter says yes/no to physical candidates for membership of the set of electrical charges. It not only identifies set-members (hard data) but also counts them. This kind of hard data collection is now so commonplace in our technological world that it perhaps induces us to regard the data problem as trivially easy to master. But in the medieval, pre-Galilean study of motions of bodies (with its idea that "motion is the realization of a body's potential"), the distinction between soft and hard data was far from obvious, and the resolution of the problem occupied many generations of scholars before a decisive breakthrough was achieved by Galileo. Many of the problems and situations that we find necessary to deal with today do not have this long history of development.

While a geographer can probably feel quite happy about the hardness of much of his data (for example, the set of streets in a town, or the set of contours over a region), he might be tempted in some situations to substitute soft data for hard, particularly in those areas where "human geography" creeps over into sociology. In the field of urban studies, how do we obtain hard data relevant to the planners asking the question, What urban structure is most conducive to making your town a pleasant place to live? Here we need a set whose members are well-defined things called "urban structures," and then we need a set whose members constitute well-defined things called "pleasant places to live." The latter seems to lead inevitably to the question of "value judgments," and where is the hard data in that context? Many social scientists would probably argue that the presence of such judgments must mean that mathematical-set-data is fundamentally irrelevant, and so its pursuit is an illusion. But this view ultimately negates the scientific approach and is not unlike the Aristotelian objections to Galileo's definition of velocity (as distance divided by time), which, apparently, removed all the "content" (or poetry) from motion in making it hard.

Where is the hard data in the replies to a questionnaire that asks, Do you think that Choice A is worse than/as good as/better than Choice B? On the one hand, this presents the citizen with a well-defined set whose members are three in number—that is, worse than, as good as, and better than. But is this set well defined among the observers (all of them) or citizens? Can any two people agree about "worse than" or "as good as"? Also, is it clear that any one choice is static? Is it not often the case that one feels the need to answer both "worse than" and "better than"? In other words, one might really need to make a selection from the *power set,* $P(X)$, of such a set X and select a *subset,* rather than a single element.

So, well-defined sets certainly require (1) a common agreement about the sets (agreements among the scientific practitioners) and (2) an appreciation of the logical (and therefore methodological) distinction between a set X and its power set $P(X)$ (and so of $P^2(X)$, $P^3(X)$, and so on). This latter point is also fundamental to the avoidance of the many set-theory paradoxes that logicians have recently brought to light. A striking instance of this was provided by Bertrand Russell in his famous Barber Paradox, which runs as follows: "In a certain town (Seville?) every male either shaves himself or else he is a shaved by the (male) barber. Does the barber shave himself?"

If we assume that the barber shaves himself, then we can deduce that he is a man who shaves himself and therefore he is not shaved by the barber. If we assume that the barber does not shave himself, then we deduce that he is a man who does not shave himself and therefore that he is shaved by the barber. We therefore obtain a logical paradox, which is illustrated by trying to list a relation λ between two sets Y and X. The set $Y = \{\text{barber}\} = \{B\}$, while the set $X = \{\text{men}\} = \{M_1, M_2, \ldots\}$. The problem arises when we try to include B in the set X itself. Thus, let λ be a simple matrix of 0's and 1's as shown:

λ	M_1	M_2	M_3	M_4	M_5	M_6	M_7	B
B	1	0	1	1	0	1	0	?

In this scheme the men M_1, M_3, M_4, M_6 are shaved by B, but if $B \in X$ (if the barber is a man), then we cannot decide on the entry under B in the top row. This set $X = \{M, B\}$ is not well defined in terms of λ. The reason is that the barber, B, is not a member of X, but is in fact a member of $P(X)$. The barber, B, qua barber, is really the subset of X—that is, $\{M_1, M_3, M_4, M_6\} \in P(X)$—and because of that we cannot ask questions of B [of members of $P(X)$] as if B were a man M_i (a member of X). This distinction, which is logically profound and of fundamental significance in any scientific methodology based on hard data, was classified by Russell in his *Theory of Types,* in which he insisted that we must not confuse "elements" with "sets of elements" with "sets of sets of elements" with . . . or X with $P(X)$ with $P^2(X)$ with . . .

It follows that the methodology we are seeking must depend on a hierarchical arrangement of data corresponding to these types, and the hierarchy, H, which I propose is one based on the quite general notion of mathematical cover sets. We can denote the hierarchical levels by N, $(N+1)$, $(N+2)$, and so on (or equally by $(N-2)$, $(N-1)$, N, $(N+1)$, etc.), with possible sets as in Table 4.1. We shall expect the hierarchy of data sets, H, to be defined by mathematical relations λ, μ, . . . in the following way.

If X is an N-level set and A a corresponding $(N+1)$-level set, then A must be a *cover set* for X—that is, the elements of A are subsets of X—and if

$$A = \{A_1, A_2, \cdots, A_n\}, \quad X = \{X_1, X_2, \cdots, X_m\},$$

then

$$A_i \in P(X) \quad \text{for } i = 1, \ldots, n;$$
$$X = \cup_i A_i.$$

If, in addition, we know that $A_i \cap A_j = \emptyset$ (the empty set), then we say that A is a *partition* of X; the elements of A are quite distinct. The hierarchy H is now defined by relations, like μ, by simply noticing that $A_i \, \mu \, X_j$ iff $X_j \in A_i$. This relation μ will therefore be represented by an *incidence matrix, M,* with binary entries 0/1. Naturally, the whole schema of data will contain "horizontal" relations, like λ, as well as the obvious hierarchical (vertical) relations μ, the two being necessary to allow for such diagonal (joint) relations Θ (where $\Theta = \mu \circ \lambda$).

Table 4.1. Hierarchy of Data Sets

Level	*Sets*	*Types*
$N+2$	$\cdots\cdots L \cdots\cdots$	$P^3(S)$
$N+1$	$A \cdots\cdots B \cdots\cdots$	$P^2(S)$
	μ Θ	
N	$X \;\underline{\lambda}\; Y \cdots\cdots Z \cdots\cdots$	$P(S)$
$N-1$	$P \cdots\cdots$	S

ANALYSIS OF RELATIONS $\{\lambda\}$

It follows from the preceding section that the analysis of hard data (suitably arranged in a hierarchical pattern H) depends on an analysis of a relation $\lambda \subset Y \times X$. Notice that a relation λ may be viewed as a subset of the Cartesian-produced $Y \times X$ in the sense that, if (Y_i, X_j) are λ-related, then they are in a subset of $Y \times X$, and this subset can be denoted by λ.

Such a relation λ represents what mathematicians call a *simplicial complex* (or complex of simplices), denoted by $KY(X)$, and this can be intuitively represented and appreciated as a collection of abstract convex polyhedra in a suitable Euclidean space E^N. We illustrate this in Table 4.2 with the simple relation λ.

In $KY(X)$, the X_i are the vertices; Y_i is the name of the polyhedron (tetrahedron) whose vertices are $\langle X_1\ X_2\ X_3\ X_4\rangle$; Y_2 is the name of the polyhedron (triangle) whose vertices are $\langle X_2, X_3, X_5\rangle$; and so forth. The polyhedron (the Y_i) are variously connected; we say, for example, that Y_1 and Y_2 are 1-connected because they share two vertices (these two vertices define a common edge, which is a one-dimensional connection). Generally we say Y_i and Y_j are q-connected if they share $(q + 1)$ vertices or are connected by chains of polyhedra where adjacent polyhedra share $(q + 1)$ vertices. Each polyhedron [representing a q-simplex in $KY(X)$] that possesses $(q + 1)$ vertices requires a q-dimensional subspace of E^N for its representation. If $n = \max$ (q-values in $KY(X)$), then we call n the dimension of $KY(X)$ and write $n = \dim K$. There is a well-known mathematical theorem that proves that taking $N = 2n + 1$ is sufficient to accommodate the whole of $KY(X)$ in E^N.

Naturally, the transpose of the incidence matrix of λ also represents a perspective from which to view the "data." This is called a *conjugate* complex

Table 4.2 Binary Relation Representation of Simplicial Complex

λ	X_1	X_2	X_3	X_4	X_5	X_6	X_7	X_8	X_9	X_{10}
Y_1	1	1	1	1	0	0	0	0	0	0
Y_2	0	1	1	0	1	0	0	0	0	0
Y_3	0	1	0	0	1	1	0	0	0	0
Y_4	0	0	0	0	1	1	1	1	0	0
Y_5	0	0	0	0	0	0	1	1	1	0
Y_6	0	0	0	0	0	0	0	1	1	1
Y_7	0	0	0	1	0	0	0	1	0	1
Y_8	0	0	1	1	0	0	0	0	0	1

$KX(Y)$, in which the set Y provides the vertices and X provides the names of the simplices (the polyhedra in E^M). Geometrical representations of $KY(X)$ and $KX(Y)$ for the above relation λ are given in Figures 4.1 and 4.2.

We can define a transitive relation β represented by the open statement "—— is q-connected with ——" in the following way. Suppose there exists a chain of direct connections such that Y_i is q_1-connected to Y_k, which is q_3-connected to $Y_u, \ldots,$ which is q_t-connected to Y_z. Then if q is the least of these integers $q_1, q_2, \ldots, q_t$, we say that Y_i is q-connected to Y_z. This defines the relation β; furthermore, it is easy to see that β is an equivalence relation on the Y_i [of $KY(X)$]. We can therefore find the equivalence classes at each q-level. The numbers Q_q of these are listed in the examples given in Figures 4.1 and 4.2. The list of all Q_q numbers (the result of what I have called a *Q-analysis*) is a vector, the *structure vector,* for the complex. In this special case we therefore have the following vectors:

$$\text{For } KY(X);\quad \mathbf{Q} = \{\overset{3}{2}\ \ 8\ \ 1\ \ \overset{0}{1}\};$$
$$\text{For } KX(Y);\quad \mathbf{Q} = \{\overset{3}{1}\ \ 6\ \ 2\ \ \overset{0}{1}\}.$$

It follows that a typical hierarchy H of data sets corresponds to a general structure, which is the union of complexes like $KY(X)$ and $KX(Y)$, for all the relations $\lambda, \mu, \ldots,$ in H. There is thus a structure (which is well defined) for each level N, $(N+1), \ldots,$ in H. We represent the structure at N-level by $S(N)$ and refer to it as the *static backcloth* $S(N)$ for the hard data, the word "static" being clearly relative to the data. Against such a backcloth $S(N)$, the "dynamics" associated with the system must be found. This will consist of "traffic" (for all

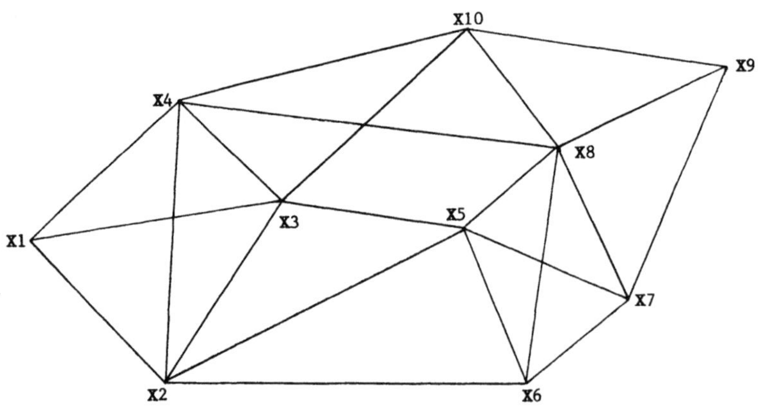

q	Q
3	2
2	8
1	1
0	1

Figure 4.1. Geometrical Representation of $KY(X)$

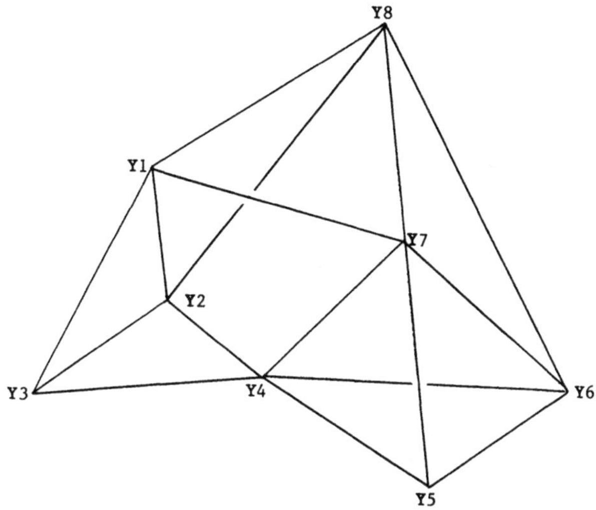

q	Q
3	1
2	6
1	2
0	1

Figure 4.2 Geometrical Representation of $KX(Y)$

things that vary in time periods that are short, relative to changes in $S(N)$ itself), except that we must specify (1) traffic on $S(N)$ [$\equiv KY(X)$] is determined by the vertex set X and (2) traffic can be represented by a *pattern* on $S(N)$, which is a mapping $\pi : Y \to J$ (J being some number system).

Studies of patterns and how changes $\delta\pi$ can be interpreted in terms of structural forces (or stresses) can be found in Atkin (1974*a,b,c*; 1975*a*). The idea is taken by analogy from the kinetics of motion. Thus, because velocity, ν, is a number (value in the real number system, R) associated with a 1-simplex (the edge that joins two vertices), it is a graded pattern

$$\nu = \pi^1 : Y \to R.$$

Changes $\delta\nu$ in this mapping correspond to, are a measure of, a *force* that acts in the structure (of all the connected points of the set Y, where the body may be found). We should want to call this force a 1-force (because $\delta\nu$ is really a $\delta\pi^1$); physicists would call it a first-order tensor. In order for the structural backcloth S (in this case, the physicists' three-dimensional space E^3) to carry this force (to allow it to be experienced), it must contain the 2-simplex (triangle) of the conventional picture. The "acceleration" is actually associated with the 2-simplex, not the edge (1-simplex) carrying $\delta\nu$ (see Figure 4.3). Because of this, if we have a pattern π^t, defined on the t-simplices of a complex K, and if it experiences an incremental change $\delta\pi^t$, we must associate that t-force with $(t + 1)$-simplices in K (provided there are any such simplices).

In a similar way the orthodox physical picture of the *torque* (a second-order tensor) illustrates our 2-force. The torque is measured by the change δh in angu-

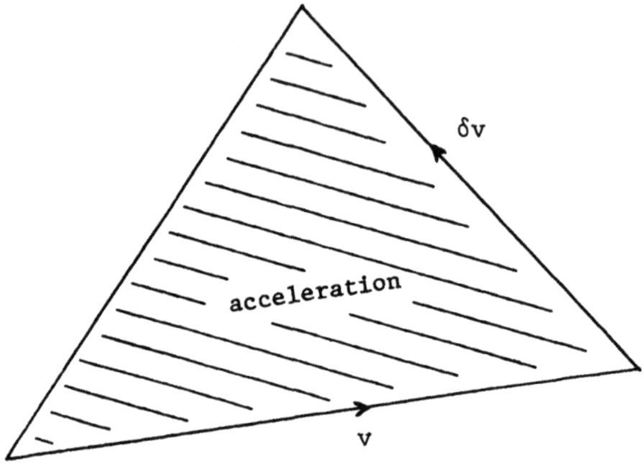

Figure 4.3. Acceleration as a 2-Simplex

lar momentum, h, the latter being associated with a piece of area (triangle or 2-simplex). Thus angular momentum is really a 2-pattern,

$$h = \pi^1 : Y \to R,$$

and the structural 2-force, δh, exists because the $KY(X)$ contains the tetrahedron (3-simplex) (see Figure 4.4).

Now these ideas can be taken over immediately into our more general backcloth, $S(N)$. If π is a graded pattern

$$\pi = \pi^0 + \pi^1 + \pi^2 + \cdots + \pi^t + \cdots + \pi^n \qquad (n = \dim K),$$

where, for example, π^t is a mapping defined on all the t-simplices of K, then we shall say that $\delta\pi^t$ measures a structural t-force in K. If σ_t is a t-simplex in K, the value of π^t on it is commonly written as (σ_t, π^t), and if $\delta\pi^t$ represents a change in π^t, we say that the structure carries the t-force (or permits it to be experienced) if there exists some $(t + 1)$-simplex, σ_{t+1}, such that σ_t is a face of σ_{t+1}. In other words, a necessary condition for a *free change* in π^t (by *free* we mean that there is no geometrical obstruction to it, that there is enough geometry in the structural backcloth to carry it) is that the various t-simplices be t-connected.

We can express this more precisely by saying that $\delta\pi^t$ shall be identified with a new pattern μ^{t+1}, defined on appropriate σ_{t+1}'s. First we take a "face operator," f, which is defined by

$$f\sigma_t = f < X_1, \ldots, X_{t+1} > = \bigcup_i < X_1, \ldots, \hat{X}_i, \ldots, X_{t+1} >,$$

where $\hat{X}_i$ means that the vertex X_i is absent. Then we demand, for example,

$$(f\sigma_t, \pi^{t-1}) = \sum_i (< X_1, \cdots, \hat{X}_i, \cdots, X_{t+1} >, \pi^{t-1})$$

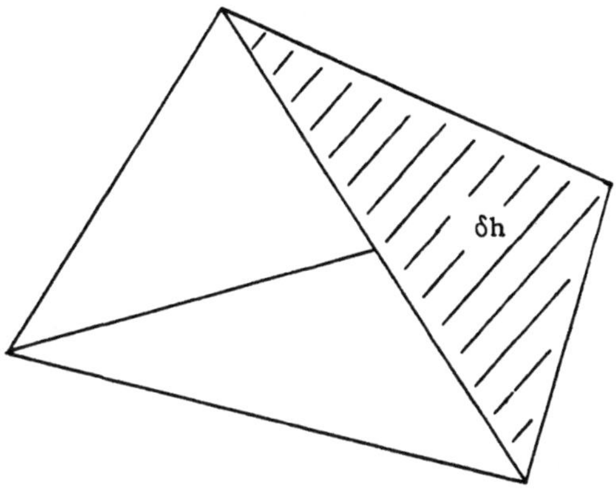

Figure 4.4. Torque as a 3-Simplex

and introduce the "coface operator" Δ by the definition

$$(f\sigma_t, \pi^{t-1}) \stackrel{\text{def}}{=} (\sigma_t, \Delta\pi^{t-1}).$$

(Notice that $\Delta\pi^t$ is a $(t+1)$-pattern). Then the μ^{t+1} required to represent the t-force $\delta\pi^t$ is identified as $\Delta\mu^t$, for some pattern μ^t, and given by making the diagram in Figure 4.5 commutative—that is, the mapping α is defined by

$$\Delta\alpha = \text{id}\ \delta = \delta \qquad (\text{id} = \text{identity}).$$

Some t-forces are "natural" to the geometry and are allowed by the connectivity structure of the backcloth $S(N)$. Some apparently are due to "outside" influences. This situation is analogous to that pertaining to the Einstein relativity theory, in which some of the forces of nature (specifically, the forces of gravitation) are due to the inherent geometry (of the four-dimensional space-time continuum), while others are not. Indeed, the use of the word "force" as describing some additional agent, brought onto the static backcloth of space-time, is peculiarly Newtonian, pre-Einsteinian. The relativity theory, in contrast, replaced this word (in the case of the gravitational force) by the idea of geometrical curvature, an idea associated with a change in the backcloth geometry [our $KC(P)$] . Indeed, as long ago as 1896, the English mathematician Clifford was making the point that a worm living in a circular tube (like the inside of a metal doughnut) would experience a permanent sense of "force" pressing it around in the circle. This experience would be exactly like our own experience of the gravitational field from which we cannot escape and which is in some obvious sense part of our sense of the geometry (the connectivity of our world). This idea was crystallized into a precise mathematical formulation in 1915 by Einstein.

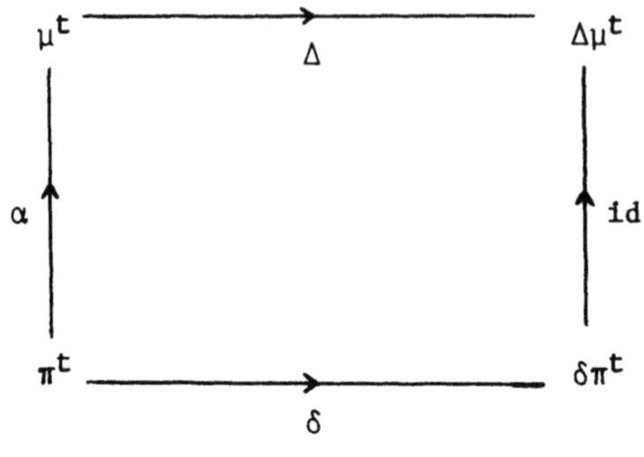

Figure 4.5. Commutativity of Operators

In the sense of our discussion to this point, we are clearly faced with this Einstein-type situation. Our idea of t-forces is dependent on the connectivity of the abstract geometry (our multidimensional and multiple-connected polyhedra) of the static backcloth $KY(X) \cup KX(Y)$. Our use of the word "force" is itself a concession to the Newtonian view, to the idea of its being due to some external agent, foreign to the given static backcloth $S(N)$. In this sense it is a concession to our intuitive experience of the geometry, of the supergeometry in which we are embedded; it is as if we are Clifford's worm, inside the tube and *not knowing it.* Knowing it is to be in the supergeometry, from which we can see the tube *from the outside.* To be inside the tube is to be Newtonian and to speak of t-forces; to be outside the tube is to be Einsteinian and to speak of the geometry as permitting or constraining the possible dynamics. These points of view are not incompatible. If, for example, the geometrical backcloth is altered by the removal of a peculiar p-simplex σ_p (perhaps by the elimination of one vertex), then any π^p defined in it must suffer a change $\Delta\pi^p$, which includes the change to zero on that σ_p. Hence this change in $S(N)$ must induce a p-force of repulsion in that part of the geometrical structure. Changes in the geometry (Einsteinian view) are therefore expressible in terms of t-forces (Newtonian view) in the original geometry.

An important feature of this approach is the realization that most of our phenomenological experience is a multidimensional one in which, generally, $n > 3$. As individuals, as citizens, as towns, in education, in medicine, in politics, all our relations lead to an existence in a multidimensional space (which is well defined in mathematical terms and therefore hard in its own right).

AN EXAMPLE OF q-CONNECTIVITY

This illustration of the use of hard data is taken from an earlier study of Southend-on-Sea (the Westcliff conservation area) and refers to commercial and visual

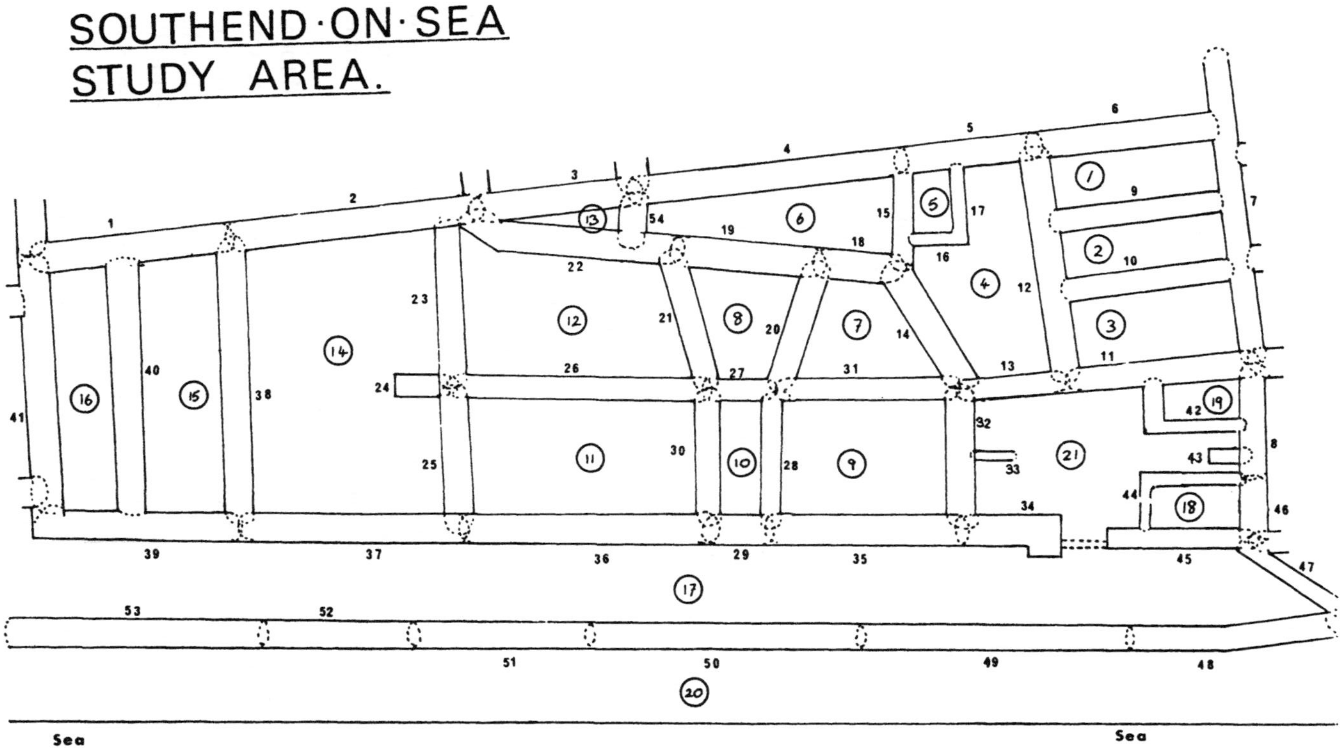

Figure 4.6. A Cover L_N for the Southend-on-Sea Study Area

effects of proposed and actual developments in certain streets of that area (see Figure 4.6). Having obtained a complex $KY(X)$, for given sets, X, Y and relation λ, one of the simplest properties of the structure is manifest via a quantity designated as the *eccentricity* of Y_i, written ecc(Y_i). This quantity measures the extent to which the polyhedron Y_i is eccentric with respect to the rest of the structure and is defined by

$$\text{ecc}(Y_i) = \frac{\hat{q} - \check{q}}{\check{q} + 1},$$

where $\hat{q}$ = top-q value of Y_i = dimension of the simplex, Y_i, and $\check{q}$ = the bottom-q value of Y_i = largest dimension of any subpolyhedron of Y_i shared with a distinct Y_j.

Generally, when Y_i has vertices not shared with any other Y_j, then it is relatively eccentric; when it shares all its vertices with another Y_j, then it is of zero eccentricity; when $q = -1$ [Y_i is totally disconnected from the rest of $KY(X)$], then ecc $(Y_i) = \infty$.

The area of our study was covered by 54 street locations in a set L_N (at the N-level), and the incidence of 152 commercial activities (a set V_N) was noted. These data were collected for 1910 and 1972. The set V_N was covered by an $(N + 1)$-level set V_{N+1} as follows (see Table 4.3): V_{N+1} = {Retail I, Retail II, Private Services, Public Services, Public Amenities, Catering, Light Industry}.

The structure $KV_{N+1}(LN)$ gave the shared-face analysis for 1910 shown in Table 4.4. Table 4.4 is read in the following way: Private Services (3) is 20-connected to Private Services (3), so it is a 20-simplex (polyhedra with 21 vertices); Private Services (3) is 18-connected to Public Services (4), so these two share a polyhedral face of 18 dimensions (19 vertices)—that is to say, they share 19 street locations (out of the 54 members of L_N), and so on.

We obtain for 1910 and 1972, in Table 4.5, comparisons of the structure vectors, **Q**, for suitable subcomplexes of $KL_N(V_n)$. Some details contained in Table 4.5 are given by the following top-q values of Table 4.6, and some eccentricities are given in Tables 4.7 and 4.8.

Under each of the $(N + 1)$-headings in Table 4.5, Retail I through Light Industry, we see that the structure vector **Q** drastically contracted between 1910 and 1972, meaning that the connectivities (in terms of shared vertices to be found in V_N) between the streets (members of L_N) became less. Details of the connectivities (Atkin, 1973) actually show that the dominant street in 1910 was $L7$ (the northern part of the High Street). This location was the simplex of highest dimension (top-q value) in each of the subcomplexes of $KL_N(V_N)$. By comparison, in Table 4.6 we see the list of simplices of *highest* dimension in 1972, in the six subcomplexes, Retail I, Retail II . . . Catering (6). Thus $L5$

Table 4.3. Relation between V_N and V_{N+1}

(1) *Retail I*	(2) *Retail II*	(3) *Private Services*	(4) *Public Services*	(5) *Community Amenity*	(6) *Catering*	(7) *Light Industry*
Butcher	Clothes (male)	Hair (male)	Gas showroom	Social society	Restaurant	Builder
Baker	Clothes (female)	Hair (female)	Art gallery	Club	Public house	Garage/sales
Greengrocer	Clothes (children)	Electric goods	Exhibition hall	Political society	Hotel	Undertaker
Fishmonger	Footwear	Radio/TV	Theater	Gardening	(Teashop)	(Boatyard)
Grocer	Ironmonger	Soft furnishings	Concert hall	Sailing	(Cafe)	
Tobacconist	Household hardware	Garage/sales	Church	Walking	(Kiosk)	
Confectioner	Electric goods	Photography	Bowling green	Fishing		
Wine merchant	Radio/TV	Betting shop	Beach sport	Water skiing		
Florist	Furniture	Chemical (drugs)	Medical/dental	Beach sport		
Off license	Soft furnishings	Bingo hall	Optometrist	Public house		
Food specialist	Garage/petrol	Builder	Post office	Library		
	Garage/car sales	Service engineer	Station (rail)	Cinema		
	Garage/access	Accounting	Public administration	Bingo hall		
	Garage/spares	Banking	Nursery	Bowling green		
	Antiques	Medical	Children's playground	Exhibition hall		
	Haberdash	Dental	Museum	Theater		
	Office equipment	Optician	Library	Concert hall		
	Jeweler	Solicitor	Special school	Youth club		
	Fancy goods	Architect	Secondary school	Amateur society		
	Photography	Surveyor	University	Flying		
	Gas showroom	Estate agent	Polytechnical school	Riding		
	Chemical (cosmetics)	Press	Art school	Golf course		
	Chemical (drugs)	Undertaker	Technical college	Bird watching		
	Chain store	Restaurant	Training college	Swimming pool		
	Bookshop	Public house	Adult education	Skating rink		
	Toys	Hotel	Youth club	Football pitch		
	Music (instruments)	Cinema	Golf course	Cricket pitch		
	Music (records)	Nursery	Swimming pool	Netball court		
	Leather goods	Auction	Skating rink	Tennis court		
	Timber/tools	Dry cleaning	Football pitch	Athletic track		
	Paint/wallpaper	Launderette	Cricket pitch			
	Art supplies	Decorator	Netball court			
	Sport shop	Finance	Tennis court			
	Newsagent	Insurance	Athletic track			
		House removal	Police station			
			Fire station			
			Hospital			
			Bus station			

Table 4.4. Shared-Face Analysis of the Structure $KV_{N+1}(L_N)$

1	*2*	*3*	*4*	*5*	*6*	*7*	*Commercial Activities*
5	0	0	–	0	0	0	1 Retail I
	12	12	10	11	6	7	2 Retail II
		20	18	19	12	7	3 Private Services
			18	18	12	6	4 Public Services
				19	12	7	5 Public Amenities
					12	4	6 Catering
						7	7 Light Industry

(Clifftown Road) shows maximum $\hat{q}$-value under Retail II ($\hat{q} = 4$) in 1972 compared with $\hat{q} = 0$ in 1910. During this period Retail I vertices moved into Clifftown Road. Similarly, the table shows the consequences of Public Services moving into Westcliff Parade from $\hat{q} = -1$ (empty) in 1910 to $\hat{q} = 2$ in 1972; Public Amenities moved into Royal Hill via the change from $\hat{q} = -1$ (empty) in 1910 to $\hat{q} = 3$ in 1972; Catering spread into Alexandra Street, Royal Terrace, High Street, and Royal Hill via the change from $\hat{q} = 0$ in 1910 to $\hat{q} = 2$ in 1972.

The final two tables, Tables 4.7 and 4.8, give us a picture of the changing role of two particular streets, Nelson Street and High Street, over this period of time. The eccentricities of Nelson Street, $L15$, in each subcomplex for 1910 and 1972 show that it has changed from being largely of low eccentricity (four values of zero) in 1910 to being largely disconnected (four values of ∞) in 1972. In 1910

Table 4.5. Structure Vectors **Q** for Suitable Subcomplexes of $KL_N(V_n)$

Subset Sliced-In	**Q** *in 1910*	**Q** *in 1972*
Retail I	$\{\overset{9}{1}\ 1\ 1\ 1\ 1\ 1\ 1\ 1\ 1\ \overset{0}{1}\}$	$\{\overset{4}{2}\ 2\ 2\ 2\ \overset{0}{2}\}$
Retail II	$\{\overset{24}{1} \cdots \overset{10}{1}\ 3\ 3\ 3\ 5\ \overset{5}{4}\ 2\ 1\ 1\ \overset{0}{1}\}$	$\{\overset{17}{1} \cdots \overset{11}{1}\ 2\ 2\ 3\ 3\ 2\ \overset{5}{2}\ 3\ \overset{3}{3}\ 3\ 4\ \overset{0}{2}\}$
Private Services	$\{\overset{30}{1} \cdots \overset{17}{1}\ 2\ 3\ 4\ \overset{1}{5}\ \overset{3}{5}\ 3\ 3\ 4\ \overset{8}{4}\ 2$ $1\ 1\ 1\ 1\ 3\ 5\ \overset{0}{2}\}$	$\{\overset{11}{1}\ 1\ 1\ 4\ \overset{7}{5}\ 5\ 6\ 4\ 3\ \overset{2}{1}\ 3\ \overset{0}{1}\}$
Public Services	$\{\overset{8}{1}\ 1\ 1\ 1\ \overset{4}{2}\ 7\ \overset{2}{10}\ 3\ \overset{0}{1}\}$	$\{\overset{2}{1}\ 5\ \overset{0}{9}\}$
Public Amenities	$\{\overset{4}{1}\ 4\ 5\ 9\ \overset{0}{3}\}$	$\{\overset{3}{1}\ 3\ 5\ \overset{0}{5}\}$
Catering	$\{\overset{3}{1}\ 1\ 2\ \overset{0}{1}\}$	$\{\overset{2}{1}\ 1\ \overset{0}{1}\}$
Light Industry	$\{\overset{5}{1}\ 1\ 1\ 1\ 4\ \overset{0}{2}\}$	$\{\overset{1}{1}\ \overset{0}{2}\}$

Table 4.6. Top-*q* Values of Table 4.5

		Street Locations	1972	1910
1.	*L*5	Clifftown Road (= *L*7)	$\hat{q} = 4$	$\hat{q} = 0$
2.	*L*7	High Street (northern end)	$\hat{q} = 17$	$\hat{q} = 24$
3.	*L*11	Alexandra Street (nos. 10–36)	$\hat{q} = 11$	$\hat{q} = 13$
4.	*L*39	Westcliff Parade (nos. 21–32)	$\hat{q} = 2$	$\hat{q} = -1$
5.	*L*47	Royal Hill	$\hat{q} = 3$	$\hat{q} = -1$
6.	*L*13	Alexandra Street (nos. 44–45/33–53)	$\hat{q} = 2$	$\hat{q} = 0$
	= *L*45	Royal Terrace	$\hat{q} = 2$	$\hat{q} = 0$
	= *L*46	High Street (southern end)	$\hat{q} = 2$	$\hat{q} = 0$
	= *L*47	Royal Hill	$\hat{q} = 2$	$\hat{q} = -1$

Table 4.7. Eccentricities of Nelsen Street (L15)

	Commercial Activities	1910	1972
1.	Retail I	0	∞
2.	Retail II	1/5	∞
3.	Private Services	2/7	2/3
4.	Public Services	0	∞
5.	Community Amenities	1	1/2
6.	Catering	0	0
7.	Light Industry	0	∞

Table 4.8. Eccentricities of High Street (*L*7)

	1910		1972	
(N + 1)	$(\check{q}, \hat{q})$	Ecc (L7)	$(\check{q}, \hat{q})$	Ecc (L7)
1.	(7, 9)	1/4	(1, 4)	3/2
2.	(8, 24)	16/9	(7, 17)	5/4
3.	(11, 30)	19/2	(4, 7)	3/5
4.	(2, 8)	2	(0, 1)	1
5.	(1, 4)	3/2	(0, 0)	0
6.	(1, 3)	1	(1, 1)	0
7.	(0, 5)	5	(–1, –1)	∞

Nelson Street was a good candidate for being a "typical street" in the structure; by 1972 it had lost all its retail trade, its light industry and its public services. Many elderly residents interviewed during the study expressed this older role of $L15$ by urging us to "have a good look at Nelson Street—it used to be the best street in town."

The final table of eccentricities for $L7$, High Street, expresses the fact that light industry disappeared from it by 1972. Even in 1910 $L7$ had its maximum eccentricity (=5) in this subcomplex—a warning sign? Although otherwise no drastic eccentricity changes have occurred, both $\check{q}$ and $\hat{q}$ values have steadily decreased. This shows that *variety has left High Street* (decrease in $\hat{q}$) and that High Street is *connected to fewer other streets* (decrease in $\check{q}$). This trend has important consequences for *traffic* (of people, goods, investment, and so on) in this area, which tends to concentrate the shopping in $L7$ at the expense of other locations.

EXAMPLE OF t-FORCES INDUCED BY BACKCLOTH CHANGES, $\Delta S(N)$

A *change* in the *structural backcloth* over some time interval, T, automatically induces changes $\delta\pi$ (and therefore structural t-forces) in any pattern π on that backcloth. The traffic that is measured by π consequently experiences these induced t-forces. For example, if Y_1 represents a 3-simplex (a tetrahedron) $\langle X_1\ X_2\ X_3\ X_4\rangle$ in some $KY(X)$ and π^3 is defined on Y_1, with value $\pi^3(Y_1)$, and during the interval T the vertex X_4 is removed from the complex, then:

1. $\langle X_1\ X_2\ X_3\ X_4\rangle$ collapses to the triangle $\langle X_1\ X_2\ X_3\rangle$.
2. $\pi^3(Y_1)$ must change to zero—that is, $\delta\pi^3 = -\pi^3(Y_1)$, because Y_1 no longer exists as a 3-simplex.
3. The traffic that π^3 measures must readjust its grading to 2 (at the most) due to the loss of the vertex X_4; it experiences a 3-force of repulsion measured by $\delta\pi^3 = -\pi^3(Y_1)$.
4. If A is a member of the original 3-traffic, it is because A has an interest in the vertices $\{X_1, X_2, X_3, X_4\}$; now A must reject X_4 as a significant vertex and become part of the 2-traffic (with interest in $\{X_1, X_2, X_3\}$). This rejection by A is the experience of the 3-force of repulsion induced by the structural change in $KY(X)$.

In a current regional study of East Anglia (Atkin 1975*b*), changes in the backcloth over the period from 1966 to 1971 have been deduced from census data, and some idea of the t-forces experienced by traffic can be seen from the results shown in Figures 4.7, 4.8, 4.9, and 4.10. In Figure 4.7 the ordinates show the

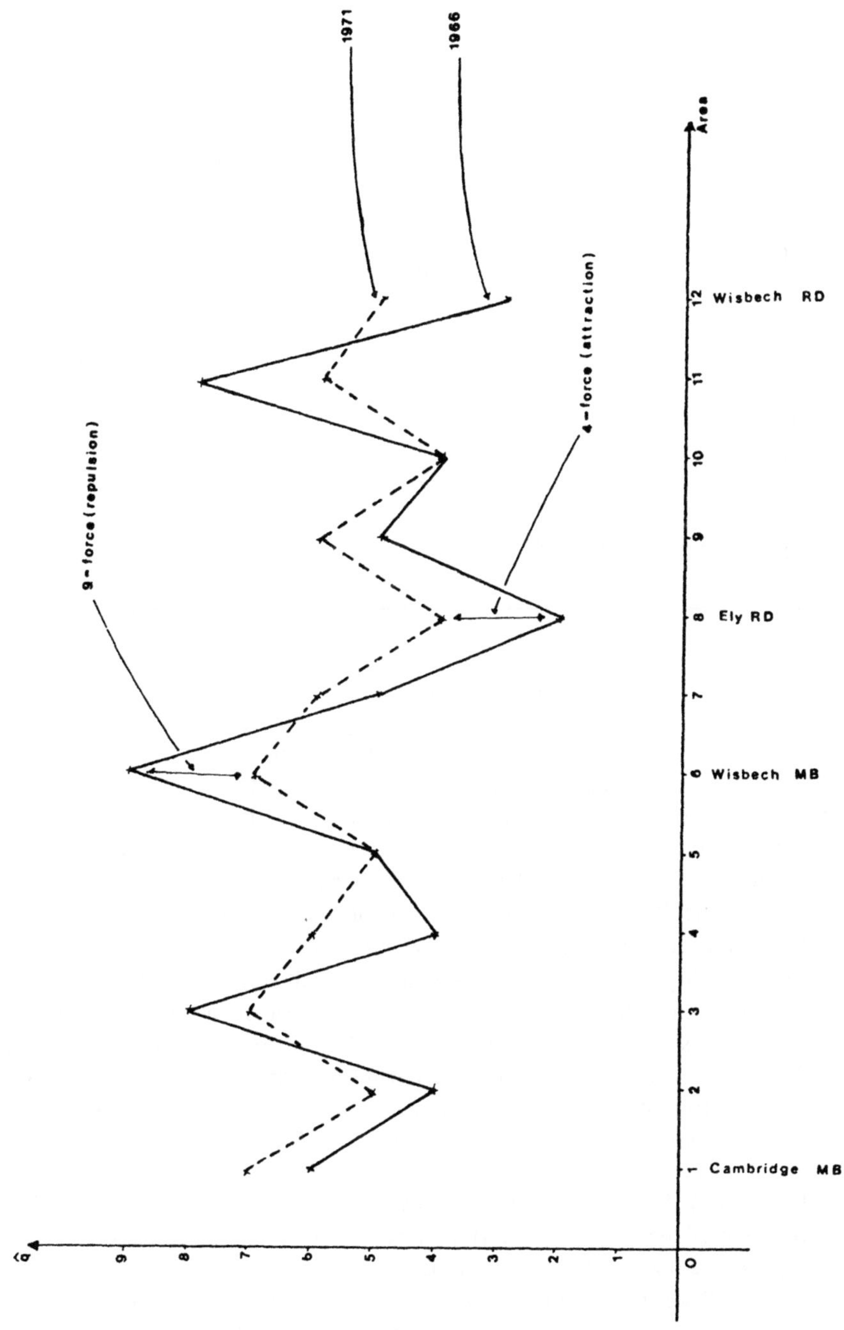

Figure 4.7. Cambridgeshire, t-Forces due to $\Delta S(N+1)$

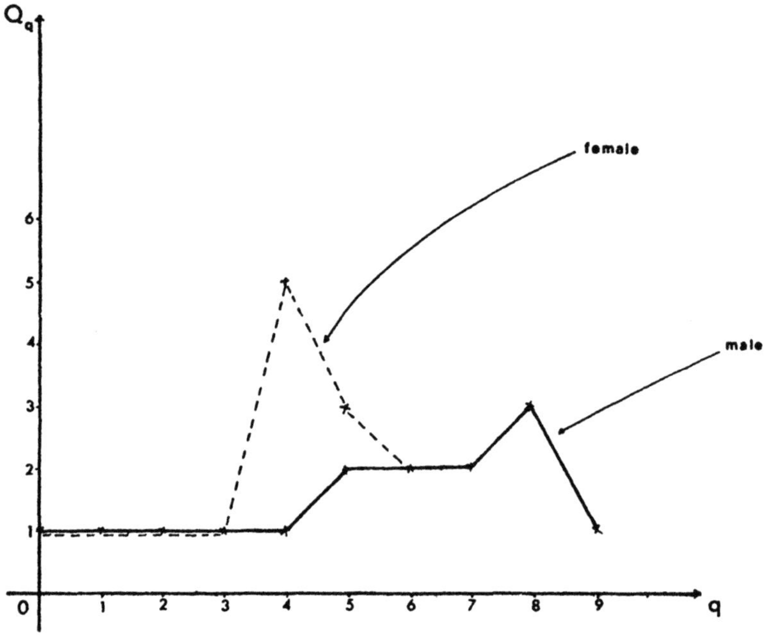

Figure 4.8. Structural Vectors, 1966 Data

top-q values of the Local Authority Areas (in Cambridgeshire, prior to the 1974 reorganization) in a complex $KY(X)$ in which the vertex set X contains the twenty-nine elements of the Standard Industrial Classification (SIC). For any one area the vertical line between its representative points on the 1966 and 1971 graphs represents the t-force that has been induced by changes in S. This t-force is attractive if the 1971 point is above the 1966 point; otherwise it is repulsive. Typical traffic that would have experienced these t-forces might be such things as classes of employment-seeking males, volume of vehicular traffic flow through each area, goods imported and exported by each area, or financial investment per area. The graphs show, for example, a 9-force of repulsion in Wisbech Municipal Borough and a 4-force of attraction in Ely Rural District. Analysis of other data shows, for example, structural forces between male-oriented and female-oriented traffic over the whole region during this period. In Figure 4.8 we illustrate the differences between the structural vectors **Q**(female) and **Q**(male) for Cambridgeshire (1966 data). In a similar way we show, in Figures 4.9 and 4.10, the top-q values for each of the areas of Cambridgeshire, for the data of 1966 and of 1971, separately for the male and female structures.

In Figure 4.9 we see that throughout the greater part of the county, there is a strong bias in favor of male-associated traffic. For example, in Wisbech MB, we

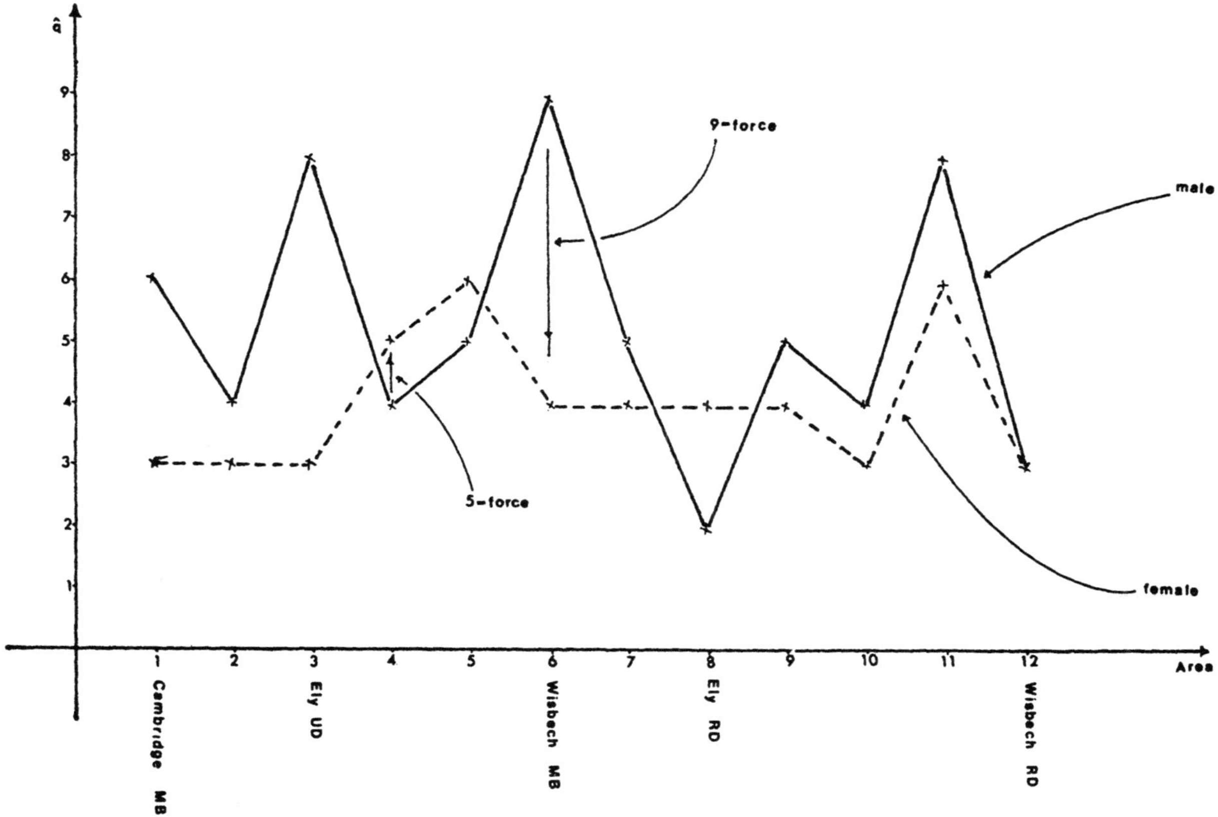

Figure 4.9. Female-Male t-Forces, 1966 Data

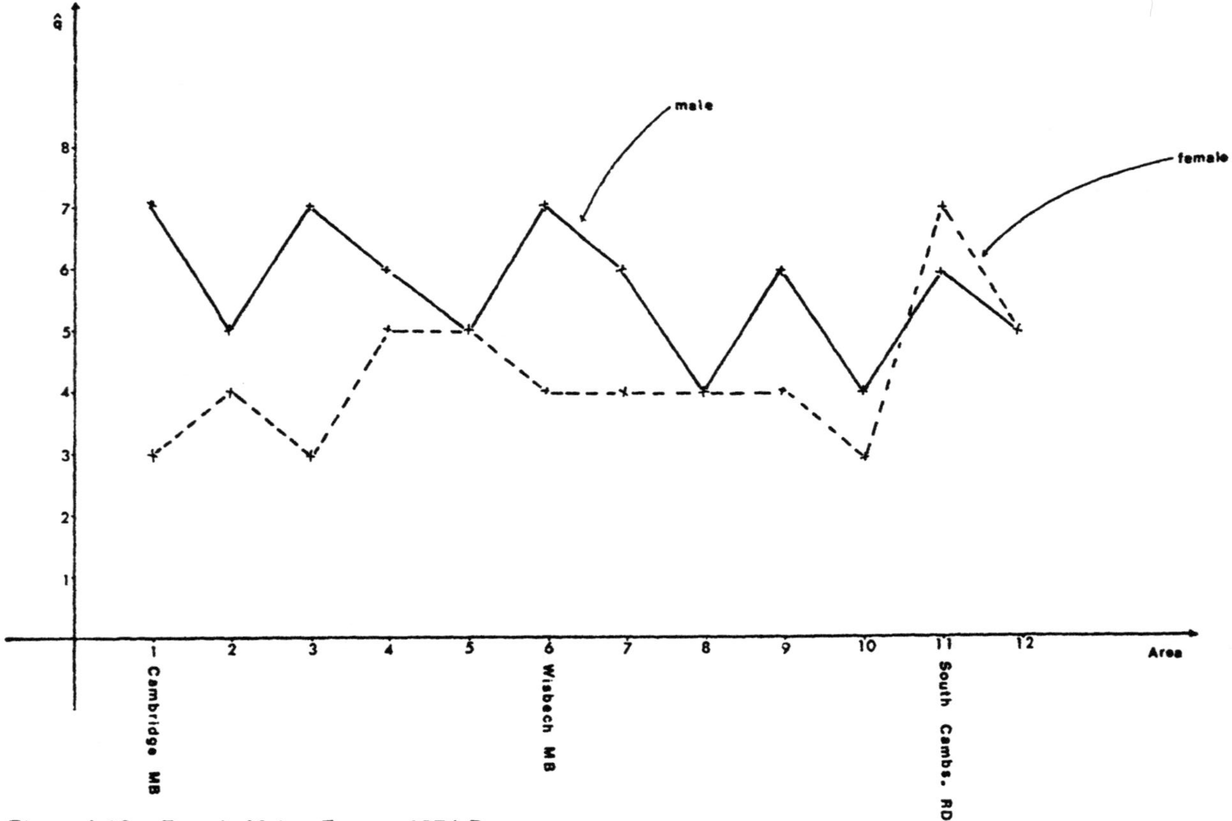

Figure 4.10. Female-Male *t*-Forces, 1971 Data

Table 4.9. t-Forces in the Female Structure $KY(X)$, 1966–1971

Area	t-force
1. Cambridge MB	Nil
2. Chatteris UD	4-force (attraction)
3. Ely UD	Nil
4. March UD	Nil
5. Whittlesley UD	6-force (repulsion)
6. Wisbech MB	Nil
7. Chesterton RD	Nil
8. Ely RD	Nil
9. Newmarket RD	Nil
10. N. Witchford RD	Nil
11. S. Cambridgeshire RD	7-force (attraction)
12. Wisbech RD	5-force (attraction)

see that the experience of shifting from the male structure to the female structure would require the traffic to overcome t-forces for which $t = 5, 6, 7, 8, 9$. The only forces of attraction on female-associated traffic occur in March UD, Whittlesley UD, and Ely RD. In Wisbech RD there is no apparent difference in the two structures.

In Figure 4.10 we see that the position has noticeably deteriorated as far as the female-associated traffic is concerned. Only in S. Cambs. RD is there any bias in favor of the female-associated traffic. In all other areas there is a total absence of positive (attractive) t-forces acting on the female-associated traffic.

The differences in the two structures $KY(X$; male) and $KY(X$; female) are quite striking. The structure (social) t-forces that must be overcome before female-associated traffic can move as freely as male-associated traffic are considerable, and they are well illustrated in Figures 4.9 and 4.10.

Table 4.9 shows the occurrence of t-forces in the female structure $KY(X$; female), over the time period 1966–1971. We see the overall effect on female-associated traffic as resulting in a repulsion from the urban district of Whittlesley, attraction to the urban district of Chatteris, attraction to the rural district of South Cambridgeshire, and attraction to the rural district of Wisbech.

DYNAMIC PATTERNS ON A STRUCTURE $S(N)$

The question of making dynamic patterns (of traffic) well defined and of relating them to a precise data-collecting process remains an important research question at this stage. But in a typical community study, we would expect the picture of dynamic traffic to involve patterns, π, which relate to variations over, say, a daily period of time (in which π is regarded as constant over any one hour), a weekly period (in which π is constant over any one day), a seasonal period of time (in which π is constant over any one week), and so forth.

Such time intervals would seem to be naturally parallel to our previous $(N-1)$-, N-, and $(N+1)$-levels, suggesting that there is a natural kind of *dynamical hierarchy*, each level of which is characterized by a time interval in which the π is constant. Then one would expect to enumerate patterns that represent, for example, at a dynamical T_1-level (where T_1 = one hour):

Pedestrian shoppers (both kinds of retail);
Traffic flow (including vehicle parking);
Catering trade (restaurants and inns);
Tourist sightseeing (by areas L_i).

Examples of patterns at a dynamical T_2-level (where $T_2 = 24T_1$) are:

Retail trade (I and II);
Private services (professional and other);
Use of leisure facilities;
Traffic flow (private, commercial);
Tourist count.

Examples of patterns at a dynamical T_3-level (where $T_3 = 7T_2$) are:

Retail trade;
Private services;
Tourist visitors;
Employment (light industry);
Employment (other).

At yet another level, and this could refer to such a period of time (say, five or ten years) that it might also be expressed as changes in the static backcloth, we would examine the patterns for:

Population distribution (by age, income, occupation);
Property values (private houses, business premises);
Building activities (public and private);
Long-term traffic control.

This latter possibility reminds us that our choice of features that go to make up the *static* nature of $S(N)$ must be a relative one. But this is allowed for in the mathematical language by the fact that every backcloth is representable by a *characteristic pattern polynomial*—say, π_0, in the algebraic representation via an exterior algebra ΛV. Change in this π_0, consequent on physical changes in the static backcloth, can then be interpreted in the dynamic language of t-forces in the inherent structure. Our overall dynamical view can therefore be given the hierarchical structure shown in Table 4.10.

We can, therefore, see that if we use our coface operator Δ, to indicate the allowed changes in a pattern (compatible with the geometry of the structure), then Δ acts as a mapping on the dynamical hierarchy in the sense that

$$\Delta : \pi(T_r) \to \pi(T_{r+1}).$$

This merely says that changes in $\pi(T_r)$ can be observed only at the T_{r+1}-level. Consequently, we need to contemplate t-forces (measured by values of $\delta\pi^t$) at any particular dynamical level. Thus a 4-force ($\delta\pi^4 \neq 0$) at the T_2-level refers to a pattern π, which is constant in the time interval T_2, and so the 4-force must refer to the next dynamical level, namely, the T_3-level. Hence we see that if π^t is a pattern at the T_r-level, then $\delta\pi^t$ is a t-force experienced at the T_{r+1}-level. This contains whatever substance there is in the temporal sense of cause and effect associated with the word "force."

In an earlier study of the Suffolk village of Lavenham (Atkin, 1975*b*) the backcloth, $S(N)$, contained the complexes $KL(V)$ and $KV(L)$, where $V \equiv V_N$ and $L \equiv L_N$ were, respectively, sets of architectural visual styles (Tudor, Georgian, and so on) and street locations. On such a backcloth we might well wish to take the total *tourist count* (over a period T_r) as a dynamic pattern π on the complex $KV(L)$. If we assume (or could ascertain) that the tourists *all* visit

Table 4.10. Dynamical Hierarchy

Time Period	*Dynamical Level*	*Pattern Polynomial*
$T_1 > 0$	T_1-level	π constant in T_1, variable in T_2
$T_2 > T_1$	T_2-level	π constant in T_2, variable in T_3
⋮	⋮	⋮
$T_n > T_{n-1}$	T_n-level	π constant in T_n, variable in T_{n+1}
⋮	⋮	⋮
T_0 (or T_∞)	Static level	π constant in T_0 (all time)

Lavenham to see the Tudor-style buildings (and nothing else), then our analysis showed that we could say $\pi \equiv \pi^{20}$.

If, on the other hand, we take a tourist count (over T_r) attributed to the individual streets in the set L, then we can regard the pattern π as a set of cosimplices on the conjugate complex $KL(V)$. This time, if there is only one attractive feature (Tudor style), we would expect that $\pi = \pi^0$, even though any particular $L_i \in L$ might well be a higher-order simplex in $KL(V)$. Such tourists, moving through the set L, see *only* the Tudor-style buildings at the N-level: any street L_k that does not contain any Tudor-style buildings (as a vertex of its simplex) will not be connected to an L_i that does (as far as our tourist is concerned). This is why changes $\delta\pi^0$ will be obstructed if the L_i are not 0-connected (sharing the single vertex Tudor). Thus the pattern π^0 would represent a set of numbers on the $L_i = \langle \ldots, \text{Tudor}, \ldots \rangle$, and these would be free to change only on sets of L_i that are connected via Tudor, in $KL(V)$.

If now we consider a set of tourists who would also appreciate the early Victorian and the Georgian styles, then the pattern π would contain the graded component π^2. Generally, therefore, we would expect such a pattern of tourist counts (on the L_i) as graded via

$$\pi = \pi^0 \oplus \pi^1 \oplus \pi^2 \oplus \pi^3 \oplus, \text{etc.}$$

Changes in the static backcloth structure $S(N)$—for example, a redevelopment that eliminated most of the late Victorian and some of the Tudor—would introduce (forcible) changes in the graded pattern π. For instance, using a typical set of vertices, we might have

$$\pi^2 = k_1 x_1 x_2 x_3 + k_2 x_2 x_3 x_4 + k_3 x_1 x_2 x_4,$$

where k_1, k_2, k_3 are some actual count on the streets-as-simplices:

$$L_i = \langle V_1 V_2 V_3 \rangle \quad L_j = \langle V_2 V_3 V_4 \rangle \quad L_k = \langle V_1 V_2 V_4 \rangle.$$

If the street L_i were to lose all its Georgian buildings (or enough of them to cause a drop below the slicing norm) and if L_k were to lose its late Victorian, we would get the forcible change $\delta\pi_0$:

$$\delta\pi_0 = \underbrace{-x_1 x_2 x_3 - x_1 x_2 x_4}_{\delta\pi_0^2} + \underbrace{x_1 x_3 + x_1 x_2}_{\delta\pi_0^1}$$

on the static backcloth. This induces a natural change in π^2 of

$$\delta\pi^2 = -k_1 x_1 x_2 x_3 - k_3 x_1 x_2 x_4,$$

which would be experienced as a *2-force of repulsion* (tending to *evacuate* the streets L_i and L_k) by the tourists (who are trying to make a contribution, by their presence, to π^2). It would also induce a *1-force of attraction*, tending to *fill* the streets $\langle V_1 V_3 \rangle$ and $\langle V_1 V_2 \rangle$. This means that the tourist obtains an increased sense of attraction, of mobility among the L_i, only if he or she is willing to live (qua a tourist) in a smaller dimension (a geometry of 1-connections). Another way of putting it is that altering the backcloth in this specific way might not decrease the total number of tourist visitors, but it will certainly alter the kind of people they are. Perhaps not unnaturally the trader might well ask, "Do π^2-people spend more money in my kind of shop than π^1-people?"

I believe that research that is to prove of practical value to all members of the community should provide an answer to this question and to similar questions. To this end the collection of data that are pattern-oriented is clearly of fundamental importance. We notice, too, that the grading of a pattern and the consequent identification of relevant t-forces are functions of the hierarchical N-level of the backcloth. Thus in the case of our hypothetical tourists who constitute a π^0-pattern (because they see only Tudor-style buildings) at the N-level, moving down to the $(N - 1)$-level (and the relevant induced relation λ), we must replace the N-word "Tudor" by the six $(N - 1)$-words that express that visual style. Suppose these form the set $\{V_1, V_2, \ldots, V_6\}$. Then it is tempting to ascribe to each of our tourists the ability to see and to search out all of these six vertices. This would make the π^0 at the N-level equivalent to π^6 at the $(N - 1)$-level.

Changes $\delta\pi^0$ on $S(N)$ are also equivalent to changes $\delta\pi^6$ at $S(N - 1)$. In terms of t-forces, we therefore see that we can experience a 6-force in $S(N - 1)$ instead of a 0-force in $S(N)$—but only if we are capable of being a "6-tourist" at the $(N - 1)$-level. It would appear that we pay for our increased sensitivity to detail by being more susceptible to higher-order t-forces. Conversely, we can blunt the effects of a t-force by moving into a position in which the backcloth $S(N)$ becomes $S(N + 1)$.

COMPARISON WITH REGRESSION ANALYSIS

These general ideas about the structure of a relation may clearly be applied to any data that can be displayed in a two-dimensional array. We can therefore compare the ideas with those that are relevant to a statistical view of the data, in particular to the method of *regression analysis* and the role of the (linear) *correlation coefficient, r.*

The idea behind regression analysis is to find a best-fit curve (in fact a straight line) between the points with coordinates (X_i, Y_j), where the data are a relation λ between the finite sets X and Y. As an illustration we shall consider the relation in Table 4.2 where X contains ten members and Y contains eight, there being a total of twenty-six points in the plane (see Figure 4.11).

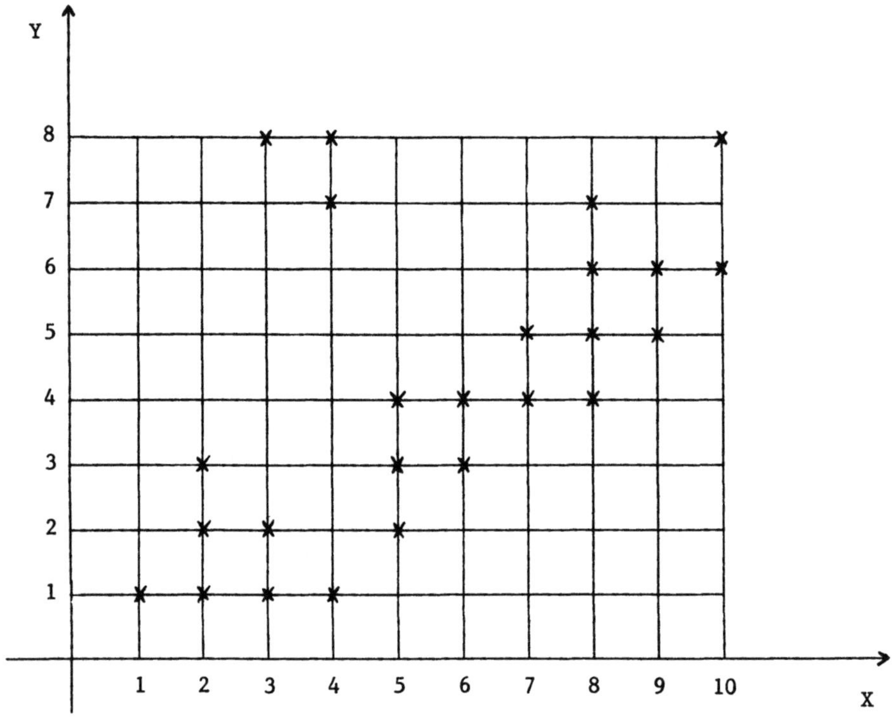

Figure 4.11. Plot of Relation λ from Table 4.2

For computational purposes we suppose that the pair (X_i, Y_j) in the relation λ possesses the coordinates (i, j). Then standard theory considers the two *regression lines*

$$Y = a + bX, \quad \text{of } Y \text{ on } X$$

and

$$X = c + dY, \quad \text{of } X \text{ on } Y$$

obtained by minimizing, for example, the expression

$$\sum_{i=1}^{N} (a + bX_i - Y_i)^2,$$

with $N = 26$.

These regression lines are shown in Figure 4.12 and have equations

$$Y = 1.5 + 0.49\,X;$$

$$X = 2.1 + 0.71Y.$$

The correlation coefficient is given by r where $r^2 = 0.49 \times 0.71 = 0.35$, so that $r = 0.59$. When the two regression lines are coincident ($r = \pm\, 1$), there is a perfect linear relation between X and Y, which means that the relation λ is actually a function (single-valued). Otherwise, r is taken as a measure of how nearly the relation λ is a linear map.

But these two regression lines fail to represent the *structure* of the complexes $KY(X)$ and $KX(Y)$, even though they attempt to replace them as follows: regression line Y on X, $Y(X)$ replaces the structure $KY(X)$, and regression line X on Y, $X(Y)$ replaces the structure $KX(Y)$. In either event we lose the multi-dimensional geometry that is inherent in the relation λ. In fact, when we accept one of these regression lines, say $Y(X)$, as a substitute for $KY(X)$, we are regard-

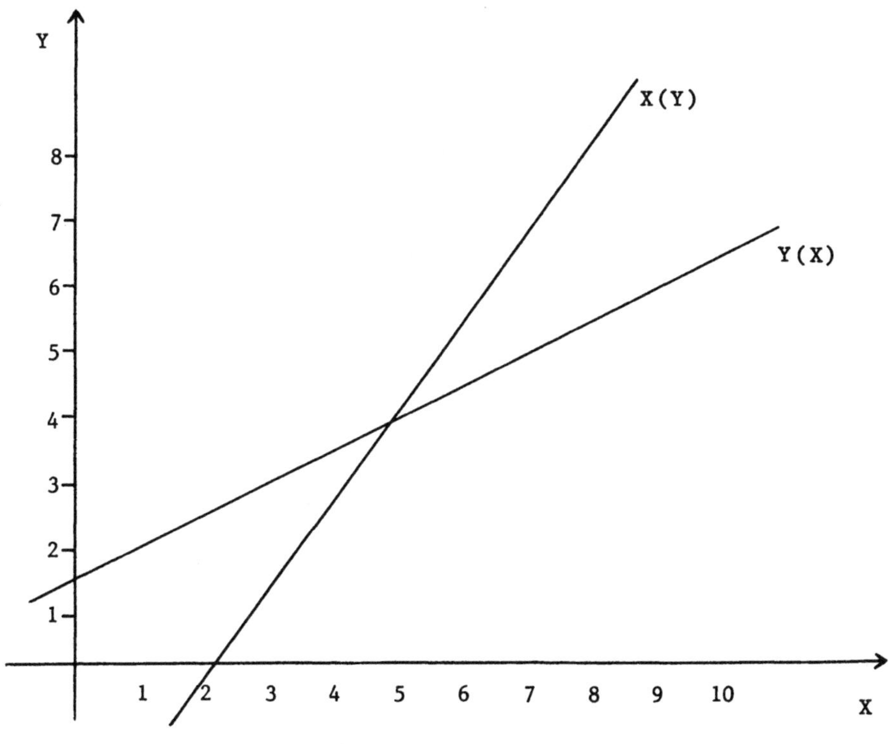

Figure 4.12. Least Squares Regression Lines for the Data of Figure 4.11

ing the structure of λ as that of a collection of disconnected 0-simplices. This follows because $Y(X)$ is now a *functional* relation between X and Y, and its incidence matrix (compare that for λ in Table 4.2) reflects this, containing only a single 1 in any one row or column. If X contains n elements, the complex $KY(X)$ is replaced by a regression-line complex whose structure vector is the simple one with $Q_0 = n$ and $Q_q = 0$ for all $q > 0$. Similarly, if Y contains m elements, the complex $KX(Y)$ is replaced by the simple structure with $Q_0 = m$ and $Q_q = 0$, for $q > 0$. The regression analysis therefore loses all the connectivity structure that is contained in the multidimensional geometry of the relation λ.

In comparison, we can attempt to compare the two structures $KY(X)$ and $KX(Y)$ directly by examining the structure vectors **Q** and **Q**$'$. By embedding these vectors in a vector space over the reals, $V(R)$, we regard this **Q**-space as a Euclidean metric space, say E^{k+1}, where $k = \max\{\dim KY(X), \dim KX(Y)\}$. If we denote an orthonormal basis for $V(R)$ by the set $\{e_0, e_1, \ldots, e_k\}$, then we have

$$\mathbf{Q} = Q_0 e_0 + Q_1 e_1 + \cdots + Q_n e_n \qquad n = \dim KY(X)$$

and

$$\mathbf{Q}' = Q'_0 e_0 + Q'_1 e_1 + \cdots + Q'_m e_m \qquad m = \dim KX(Y).$$

It then seems natural to introduce a *first structure coefficient, h,* for these complexes and define it as $\cos \Theta$, where Θ is the angle between the vectors **Q**, **Q**$'$. This will be given by the standard formula $|\mathbf{Q}| \cdot |\mathbf{Q}'| \cdot h = (\mathbf{Q}, \mathbf{Q}')$, where $|\mathbf{Q}|^2 = Q_0^2 + Q_1^2 + \ldots + Q_n^2$, and the scalar product $(\mathbf{Q}, \mathbf{Q}') = Q_0 Q'_0 + Q_1 Q'_1 + \ldots + Q_n Q'_n$, it being understood that $n \geqslant m$ and that $Q'_r = 0$ if $r > m$.

For the numerical example of this section, we have

$$\mathbf{Q} = e_0 + e_1 + 8e_2 + 2e_3;$$
$$\mathbf{Q}' = e_0 + 2e_1 + 6e_2 + e_3,$$

and so

$$|\mathbf{Q}|^2 = 72, |\mathbf{Q}'|^2 = 42 \text{ and } (\mathbf{Q}, \mathbf{Q}') = 53.$$

This gives $h = 0.96$, suggesting a much higher structural correlation than regression-line correlation.

Another interesting example of the difference between h and r occurs in the relation of Figure 4.13. Here we suppose that Y contains one element that is related to the N elements in X. Since the relation is already linear, we can take $r = 1$. But from the point of the structures, we have

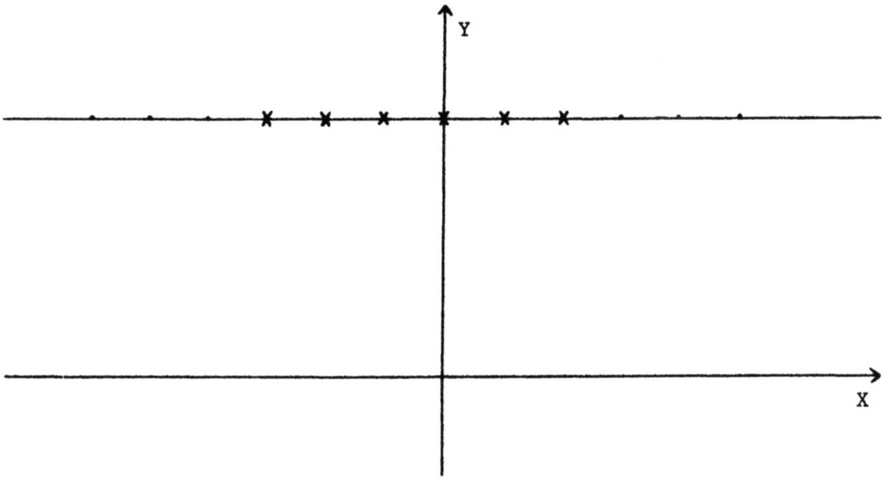

Figure 4.13. Constant Relation

$$KY(X) = \{\sigma_{N-1}, \text{only}\},\ 1 = Q_{N-1} = \cdots = Q_0;$$
$$KX(Y) = \{\sigma_0, \text{only}\},\ Q_0 = 1,\ Q_i = 0 \text{ for } i > 0.$$

We therefore get $|\mathbf{Q}|^2 = N$, $|\mathbf{Q}'|^2 = 1$ and $(\mathbf{Q}, \mathbf{Q}') = 1$ so that

$$h = 1/N^{1/2},$$

illustrating that h is a measure of structural dependence (in a multidimensional geometry). Its value corresponds closely to the intuitive idea that $KX(Y)$ is only "$1/N$ th," as structural as is $KY(X)$.

On the other hand, we would expect the structure coefficient h to give us a value of unity when the structures are identical. This will occur when the incidence matrix λ is symmetrical, so that it equals its transpose. This certainly occurs when λ defines a finite projective plane or represents a symmetric block design. From this point of view, we have the intriguing idea that in the family of all simplicial complexes, the finite projective plane corresponds to the same thing as does the linear relation in the family of all statistical scatter diagrams.

REFERENCES

Atkin, R.H. 1973. "Research Report III – A Study Area in Southend-on-Sea." University of Essex, Urban Structure Research Project (SSRC).

——. 1974*a*. *Mathematical Structure in Human Affairs*. London: Heinemann.
——. 1974*b*. "An Approach to Structure in Architectural and Urban Design, 1." *Environment and Planning B* 1:51–67.
——. 1974*c*. "An Approach to Structure in Architectural and Urban Design, 2." *Environment and Planning B* 1:173–91.
——. 1975*a*. "An Approach to Structure in Architectural and Urban Design, 3." *Environment and Planning B* 2:21–57.
——. 1975*b*. "Research Report V – A Study of East Anglia, I." University of Essex, Regional Research Project (SSRC).

5 GENERAL SYSTEMS MODELING OF CONFLICT WITHIN NATIONS

Roger Cavallo and Eduard Ziegenhagen

EMPIRICAL STUDIES OF CONFLICT WITHIN NATIONS

The study of varying forms of conflict within nations has long been an important area of focus from within political science. The orientation of much of this research has been on intensive study of single aspects of individual cases or on comparison of one of these aspects – such as revolution – over a small number of cases. In the last decade or so, an alternate orientation has developed that attempts to take advantage of modeling forms that allow the processing of larger quantities of data than had been previously possible. The main focus of this orientation has been on cross-national study of relations among various domestic conflict indicators (e.g., Rummel, 1963, 1965; Tanter, 1966; Firestone and McCormick, 1972; Banks, 1972; Gurr and Bishop, 1976; Hibbs, 1973), with the intention of constructing verifiable general statements about conflict activity.

Research initiated by Rummel reflects an example of the desire to mediate between, on the one hand, in-depth verbal and interpretive studies of individual cases and, on the other, grand and sweeping theories of why nations behave the

Part of the material presented in this chapter has been described in Cavallo (1979) but is repeated here so that the chapter represents a relatively self-contained illustration.

way they do – theories that are often substantiated by empirical evidence that is hardly more than anecdotal. Later attempts represent an effort to proceed in a manner that represents systematic integration of more than a few instances of phenomena under consideration – that is, to proceed in a manner that uses systematic methods for the ordering and surveying of human experience.

All of the research in this category utilizes data drawn from similar sources – for example, newspaper accounts, *Facts on File, Deadline Data on World Affairs* – and fits into the category classified as *events research* (Azar and Ben-Dak, 1975). The compilation of such data, and its availability for further research, represents an achievement in its own right as a first step toward a more complete understanding of political phenomena. This nonexperimental research area is thus one in which the tension between theoretical and empirical orientations especially highlights the need for an operational investigative framework.

Essentially all of the past research in this area has centered around the linear system-based statistical tools of correlation analysis, regression analysis, and factor analysis. Indeed, many social scientists – while often recognizing the unsatisfactory nature of the situation (Singer, 1976) – consider these tools as constituting the only techniques available to workers in this field.

Over the years Rummel has probably been the most enthusiastic proponent of factor analysis (e.g., Rummel, 1970), and it is interesting to consider his investigations from the perspective described in Chapter 1. Rummel indicates a basic affinity with systemic emphases, arguing:

> With its accent on mathematics and the interdependencies of elements within a system, general systems theory has been a stimulus to social theorizing. It changed the focus from phenomena to patterns and relations, and has had a purging and heuristic effect on current social thought and research. [Rummel, 1965, p. 184]

Rummel goes on, however, to argue that general systems theory has stagnated for two intricately related reasons. The first of these is that, "as used by social scientists," general systems theory has not gone beyond the provision of a conceptual and "verbal edifice." In this regard Rummel possibly underemphasizes the importance of the shifted emphases that general systems and cybernetic foci have represented (see, for example, Deutsch, 1963). His second argument, however – that general systems theory was "ungrounded in empirical data and operational concerns" and that it did not provide "a developed set of empirical methods" – relates directly to the issues that have motivated the methodological approaches described in this book and, in particular, GSPS.

With respect to Rummel's two arguments, the lack of empirical methods – or, more accurately, the failure to relate empirically significant ideas and methods to social science research – has probably been responsible for the primarily conceptual utilization of general systems research by social scientists. In this regard we recognize that "conceptual" refers to both purely verbal theories and to purely symbolic or mathematical explications that do not give consideration to investigative utility. Rummel has thus been led to total reliance on product moment correlation, multiple regression, and principle axes factor analysis to supply a model with which to deal with empirical relations. This model has subsequently been used to formulate a "field theory" of social action, which is essentially a system of linear equations deriving from a basic "axiom" that the relation between two nations that defines their behavior is a "linear vector function" of attributes that the nations possess. Rummel, in fact, recognized the restrictiveness of the assumptions involved in the use of his models but evidently felt that there was no alternative. A major purpose of the framework described here is to extend this methodological scope.

A major orientation of past studies of domestic conflict has been the attempt to determine, through the use of factor analysis, a number of "dimensions" that is lower than the number of attributes by which the conflict system is defined, such that these dimensions adequately explain the overall system. "Adequate explanation" must be defined in terms of the approach taken and in this context refers to accounting for an acceptably high proportion of the overall variance. In terms of GSPS this goal can be considered as the determination of a structure system that is derived from the overall conflict situation. Thus, a number of systems problems were implicitly involved in these past studies. The use of GSPS would involve explicit statement of all the problems as well as an attempt to use general procedures in their solution rather than those that impose structure on the problem, as is the case with factor analysis. An exposition of this past work, as well as a more extensive set of references, is included in Cavallo (1979). To illustrate the form of results those studies generated, we present in Figure 5.1 the factor structure determined in one of the studies (Banks, 1972) for conflict measures: Assassinations, Guerrilla Warfare, Major Government Crises, Purges, Riots, Revolutions, and Anti-Government Demonstrations. While factor analyses do not usually utilize the interpretation of structure and subsystems,

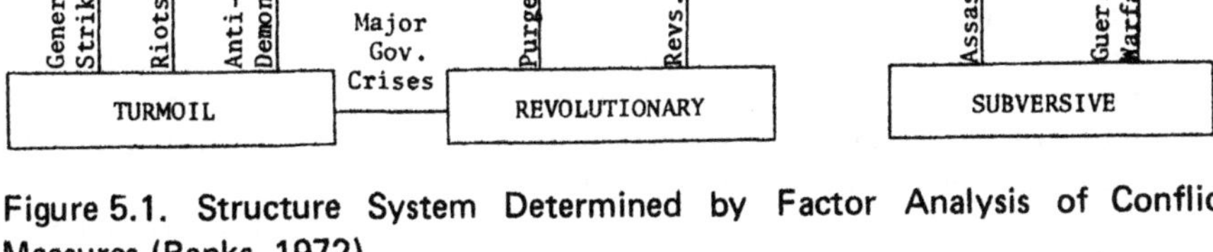

Figure 5.1. Structure System Determined by Factor Analysis of Conflict Measures (Banks, 1972)

such interpretation is natural in this context and is reflected in the block diagram form of Figure 5.1.

It is important to recognize that no working algorithm or general methodological procedure can eliminate the need for interpretation of results in context. Utilizing as many meaningful and integrated procedures as possible is thus all the more important. Consideration of past results emphasizes the desirability of a reconsideration of the data from this integrated and more comprehensive methodological perspective. In the sequel we utilize data compiled by Banks at the Center for Comparative Political Research at the State University of New York at Binghamton. We also emphasize that the procedures we utilize consider each data entry as a separate point and do not involve the loss of information that is inherent in the procedure used in all of the above studies of summing observations over a period of years. Such a procedure introduces distortion of actual interaction patterns, especially with respect to attempts to determine dynamics involved in the overall system.

SOURCE SYSTEM AND DATA SYSTEM

In the overall investigation described here, the object is thus "conflict within nations." Definition of the source system requires the specification of attributes and a set of appearances for each attribute. In this case the basic attributes have been determined in consideration of the past studies, and we decided to utilize the set consisting of six elements: General Strikes, Major Government Crises, Purges, Riots, Revolutions, Anti-Government Demonstrations. Assassinations and Guerrilla Warfare were not used, since appearances of these attributes are so highly susceptible to nonsystemic effects. Definition of the six attributes are, as given by Rummel (1963, p. 5):

Number of general strikes: any strike of 1,000 or more industrial or service workers that involves more than one employer and that is aimed at national government policies or authority.

Number of major government crises: any rapidly developing situation that threatens to bring the downfall of the present regime—excluding situations of revolt aimed at such an overthrow.

Number of purges: any systematic elimination by jailing or execution of political opposition within the ranks of the regime or the opposition.

Number of riots: any violent demonstration or clash of more than 100 citizens involving the use of physical force.

Number of revolutions: any illegal or forced change in the top governmental elite, any attempt at such a change, or any successful or unsuccessful armed rebellion whose aim is independence from the central government.

Number of anti-government demonstrations: any peaceful public gathering of at least 100 people for the primary purpose of displaying or voicing their opposition to government policies or authority, excluding those demonstrations of a distinctly anti-foreign nature.

The appearances of each attribute are the recorded events as given in the above-named data bank. For the definition of the supporting attributes and their appearance sets, each of the fifty-one countries used in the Banks study was used with observations taken on each country for the post-World War II period consisting of the years 1946–1975.

The abstraction of these attributes and their appearances (definition of a general image system) requires a decision involving transformations from the sets of appearances to the sets of recognized potential states of abstract variables. With respect to this decision, we had to take into account two basic considerations. These were (1) the limitations inherent in cross-sectional data, which force simultaneous consideration of data collected in diverse settings and (2) the nature and complexity of the overall situation, coupled with difficulties associated with the collection of data, which make utilization of a low-resolution level and search primarily for patterns that the data exhibit more meaningful. In these regards the fact that a given event has different significance in different social and cultural settings was taken into account by associating each attribute with a binary variable, where the state was determined by whether or not the appearance of the country in question was above or below the mean for that country over the thirty-year period. In this manner the state of the variable reflects the relative weight of events in the context history of each nation rather than in comparison to the frequency of events in all nations. The data system was thus represented by a data array consisting of a 6 × 1530 binary matrix with delimiters corresponding to countries set at thirty-column intervals. Table 5.1 lists the fifty-one countries used in the study.

A complete set of the data associated with each country (both raw data and transformed data) is included in Cavallo (1979). To illustrate the determination of the general image system, we here reproduce one section of the overall data array (the portion representing South Africa, Figure 5.2*b*) and the original data from which it was derived (Figure 5.2*a*). The rows of the matrices in Figure 5.2 represent the attributes and variables in the order previously given, that is:

Row 1 : v_1 : General Strikes;
Row 2 : v_2 : Major Government Crises;
Row 3 : v_3 : Purges;
Row 4 : v_4 : Riots;
Row 5 : v_5 : Revolutions;
Row 6 : v_6 : Anti-Government Demonstrations.

Table 5.1. Fifty-One Countries in Continuous Existence since 1918

Afghanistan	Canada	Honduras
Bolivia	Albania	Iran
South Africa	Chile	Italy
Spain	China	Hungary
Norway	Colombia	Liberia
Sweden	Costa Rica	Luxembourg
Switzerland	Cuba	Belgium
Thailand	Denmark	Mexico
Turkey	Ecuador	Netherlands
Soviet Union	El Salvador	New Zealand
Brazil	Finland	Nicaragua
United Kingdom	France	Panama
United States	Argentina	Paraguay
Uruguay	Greece	Peru
Venezuela	Guatemala	Portugal
Yugoslavia	Haiti	Rumania
Bulgaria	Australia	Saudi Arabia

Raw Data for South Africa (Rows, as given on p. 79; Columns = 1946–1975):

2	1	0	0	0	1	0	0	0	0	0	0	0	0	0	0	0	0	0	0	0	0	0	0	0	0	1	1	0	0
0	0	0	0	0	0	1	0	0	0	0	0	0	0	1	0	0	0	0	0	0	0	0	1	2	0	0	0	0	0
1	0	0	1	1	0	2	1	0	0	1	0	1	0	1	0	2	1	0	1	0	0	0	0	0	3	2	0	0	2
0	0	0	5	3	1	14	2	3	0	2	2	1	2	8	3	1	2	0	0	0	2	0	0	2	0	5	2	4	0
0	0	0	0	0	0	0	0	0	0	0	0	0	0	0	0	0	0	0	0	0	0	0	0	0	0	0	0	0	0
2	0	0	1	0	0	9	1	0	1	1	3	3	1	2	0	3	0	0	0	0	0	0	1	3	0	3	0	0	0

a

Columns 61–90 of Data Array Representing Data System:

1	1	0	0	0	1	0	0	0	0	0	0	0	0	0	0	0	0	0	0	0	0	0	0	0	0	1	1	0	0
0	0	0	0	0	0	1	0	0	0	0	0	0	0	1	0	0	0	0	0	0	0	0	1	1	0	0	0	0	0
1	0	0	1	1	0	1	1	0	0	1	0	1	0	1	0	1	1	0	1	0	0	0	0	0	1	1	0	0	1
0	0	0	1	1	0	1	0	1	0	0	0	0	0	1	1	0	0	0	0	0	0	0	0	0	0	1	0	1	0
0	0	0	0	0	0	0	0	0	0	0	0	0	0	0	0	0	0	0	0	0	0	0	0	0	0	0	0	0	0
1	0	0	0	0	0	1	0	0	0	0	1	1	0	1	0	1	0	0	0	0	0	0	0	1	0	1	0	0	0

b

Figure 5.2. Transformation from Object System to General Image System

As an indication of the effect of transformation to general image system, observe that the eighth column of Figure 5.2*a* indicates one instance of Purges, two Riots, and one Anti-Government Demonstration reported for South Africa for the year 1953. Since the two Riots and one Anti-Government Demonstration are lower than the average over the thirty-year period, the respective places in the eighth column of Figure 5.2*b* are assigned zeros, while evidence of the Purges is retained.

BEHAVIOR SYSTEMS

The first class of problems that are relevant in the context of GSPS as described in Chapter 1 relates to the determination of behavior systems from data systems. The essence of the methodological perspective of GSPS motivates the specification of different problems and thus the determination of different behavior systems to answer different questions that arise directly from the consideration of the specific system. The most direct problem is the determination of the memoryless behavior from the data array—that is, the specification of the behavior representing interaction among only current values of the basic variables. Formulated as a problem in GSPS, this situation is represented by:

1. A given particular initial (data) system;
2. A given terminal system type (behavior);
3. Constraints, specified as the request for memoryless behavior.

The solution that we utilize here consists of a list of the samples that occur, along with the frequency-defined probability of each sample. For example, the numeral 010101, representing one of the sixty-four possible states of this behavior system, has an associated probability of .009, indicating that over the whole data array the particular combination—Major Government Crises, Riots, and Anti-Government Demonstrations—represented roughly 1 percent of the sample. We observe that the list so generated represents information on all occurrences and interactions within each nation over the full thirty-year period. Investigative areas such as we are describing are especially characterized by the lack of a well-defined and accepted set of general ordering principles. In these cases maximum utilization of general systems methodological research implies interpretive interaction with the user, involving exploratory investigations that can fruitfully be made at all levels.

For example, even though the behavior system may not have represented a primary original investigative goal with respect to the object-system, its determination will, in general, be worth examining for the detection of patterns or interactions of interest. Independent of context and interpretation, a list of samples with probability distributions will, in general, have little significance, but it nevertheless represents a codification in a more accessible form of all the data that the context provides.

As the framework involves the association of methodological procedures—such as clustering techniques, pattern recognition algorithms, and fuzzy algorithms—to various systems problems, interpretation and interaction with context often naturally suggest systems problems that involve these procedures. One example of a generally important scientific and systems problem that is also relevant in the context of this investigation is the determination of idiosyncratic or anomalous system states. The problem is worth describing in some detail since it offers a good example of the general methodological approach. The detection of such anomalies is especially useful in the investigation of complex areas for which theoretical formulations—if they are possible—do not exist. Such is particularly the case with past political conflict modeling, which tends to obscure the identity of complex conflict events for the sake of parsimony.

A characterization of states as anomalous can be made from different perspectives. One of the simplest but most important system principles is based on the recognition that groups of phenomena often exhibit properties that cannot be predicted from observation of the individual phenomena in isolation, or, similarly, that the behavior of various system attributes in interaction is often other than expected from observation of any subset of the attributes in isolation. It is thus natural to combine this principle with a recognition of the potential significance of the detection of anomalous situations. This leads us to ask if any of the system states (indicating combinations of occurrence and nonoccurrence of each of the forms of conflict) possesses an unexpectedly high or low frequency of occurrence. Here, "unexpectedly high or low" can be meaningfully defined as a frequency that is significantly different from that associated with states (combinations of conflict) that are similar but not identical to the given state. We observe that it would be difficult in terms of this problem and the potentially extremely high number of states to give a context-independent definition of "significantly different," just as it would be difficult to check each state by singular and independent observation.

The value of the systems-methodological orientation here becomes apparent as this anomalous-state-detection problem can be easily reformulated in terms of a well-developed specific approach to systems modeling, that of an interconnected net of individual cells where the state of a given cell is determined concurrently by the states of neighboring cells. The approach has been used fruitfully in various contexts—for example, to model the engulfment of one form of tissue by another (Rogers, 1977)—and one development that is relevant to the problem discussed here is work done by Barto and Davis (1978) on the development of algorithms to facilitate calculations in consensus-competition models.

Their algorithms deal with the case where each cell is presumed to be weighted by some function with one of a range of values (e.g., -1, 0, 1) assigned to each cell in accordance with the weight of the cell. The consensus-competition descriptions then model the process whereby each cell maintains or changes its assigned value depending on the strength of the weights that its neighboring cell possesses. While the description of the algorithm is somewhat complicated, the major point of interest here is that the abstract operation of the algorithm is irrelevant to the problem and interpretation as we have described it. That is, the problem is primarily, if not solely, a data-processing problem, and the "validity" of any results—once they are determined—can be checked independently of the means utilized in their determination. This point is basic to the issues described in Chapter 1 and is dependent only on the development and use of *general* methods, which are, in fact, derived from and adaptable to problems rather than beginning with "powerful" procedures and looking for places in which to utilize them.

The consensus-competition algorithms are perfectly suited to the anomalous state problem if each state of the conflict system is considered to be a "cell" in the net where the neighbors of each cell (those states to which it is connected) are those that differ in only one of the six binary (0 or 1) digits representing the state. The function that weights each of the cells (states) is most naturally chosen to be the frequency-derived probability distribution function.

When the algorithm was used on this particular "net," all states but four maintained their originally assigned values—that is, the weights associated with each state were "reasonably" close to those of its neighbors. The process thus detected four "anomalies." This immediately suggests two further activities in context of the overall investigation: (1) to interpret each of the four abstractly determined anomalies to see if these can be explained in terms of context and (2) to attempt to discover a more complex "explanation" for those whose explanations are not immediately obvious.

With respect to (1) above, the anomalous nature of three of the four states was easily explainable in terms of the interpretation of the states. The occurring conflict activities for these three states were:

State (100100): General Strikes and Riots;
State (101000): General Strikes and Purges;
State (110000): General Strikes and Major Government Crises.

For each of these three states, the data processing indicated a change in value from low to high, indicating that each of the states appeared significantly less over the whole process than a weighted combination of its neighbors. This indication is reasonable and explainable by the fact that one neighbor of each of the three states was the state representing General Strikes as the only conflict activity. As is reasonable to expect, the tendency in the data was that, in general, the higher the number of conflict activities represented in a particular state, the lower the likelihood of occurrence of that state. In this case the relatively high occurrence of the state representing only Strike activity explains the anomalous characterization of each of the three states.

The fourth anomalous state, however, is interesting in that it involves a large amount of conflict activity and occurs significantly *more* often than its neighboring states. The state (011110) was that indicating the co-occurrence of Major Government Crises, Purges, Riots, and Revolutionary Activity. This state occurred fifteen times. Two of its neighbors with their associated frequencies given in parentheses were, for example, state (010110): Government Crises, Riots, Revolutions (4); state (001110): Purges, Riots, Revolutions (4). The higher incidence of Riots and Revolutionary activity in the presence of combined Crisis and Purge activity, compared to that in the presence of Crises or Purge activity alone, indicates an interesting phenomenon with respect to general studies that have been made in this field.

In the context of a large literature devoted to the study of individual aspects of domestic conflict, consideration of the frequency of this high conflict state suggests the need to provide a more elaborate and systematic specification of linkages among domestic conflict events than has generally been utilized. This is especially important in light of attempts to extend consideration of conflict events to formulations related to policy questions. Development of this specification and its implications for various issues requires an extensive consideration of past studies in this area; in the following paragraphs we outline some basic issues.

The orientation of many earlier works has been to consider only one aspect of apparently important relationships that are operatory in the overall social context within which policy must be formulated. This aspect has to do with the effect of dominant factors associated with social, economic, political, and psychological aspects of the nonconflict social context on particular conflict activities. In these formulations the dominant factors are considered and described as precursors of particular conflict events.

For example, economic prosperity is argued to be a precursor of revolution by Brinton (1956), Edwards (1970), and Pettee (1938). Davies (1963) and Tanter (1966) argue that deterioration preceding prosperity is the precursor of revolution. Domestic conflict in the narrower sense of Riots is attributed to social disorganization (Kornhauser, 1959; Downes, 1968), political structure (Lupsha, 1969), deprivation (Downes, 1968), and relative deprivation (Gurr, 1968; Spiegel, 1971). Such specifications of the preconditions for domestic conflict have been constructed on conceptions of domestic conflict events as functions of the underlying aspects of a dominant system. All of the forms of domestic conflict events have been considered primarily as epiphenomena to be

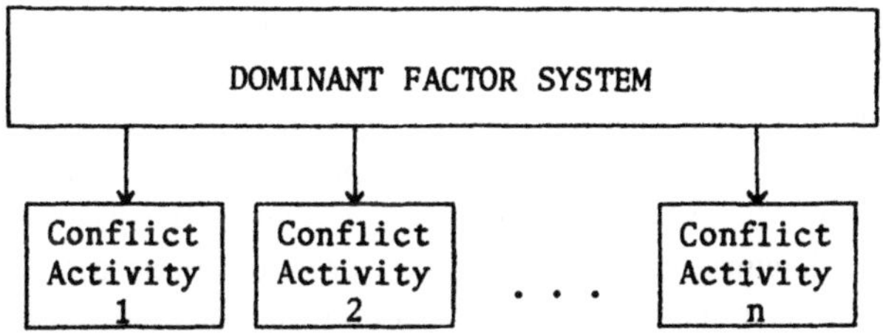

Figure 5.3. Simple Model: Linkages of Dominant Factor System to Conflict Activities

attributed to dominant-factor-system dysfunction of some variety. Interaction among conflict activities is attributable directly to the effect of the dominant system factors. The model involved in this characterization may be visualized as in Figure 5.3.

A somewhat extended conceptualization associated with this theme has emerged from the empirical work of Rummel (1963, 1965), Tanter (1966), and Banks (1972). A sense of relationship among conflict indicators is expressed in their attempts to identify a structure within the system of conflict events. These researchers have observed the persistence of various combinations of domestic conflict events among nations and across time and have related these events to more general considerations. Banks (1972), Hibbs (1973), Terrell (1971), and Feierabend and Feierabend (1966) link domestic conflict dimensions to public policies and social structure. Although providing the basis for a more elaborate conceptualization than that represented in Figure 5.3, there is still a primary concern with the influence of the dominant factor aspect. We may represent this model as in Figure 5.4, where various forms of conflict activity may now be considered to influence each other, although these influences are generally thought of as indirect and are considered to occur primarily through linkages relating individual conflict activities to the dominant factors.

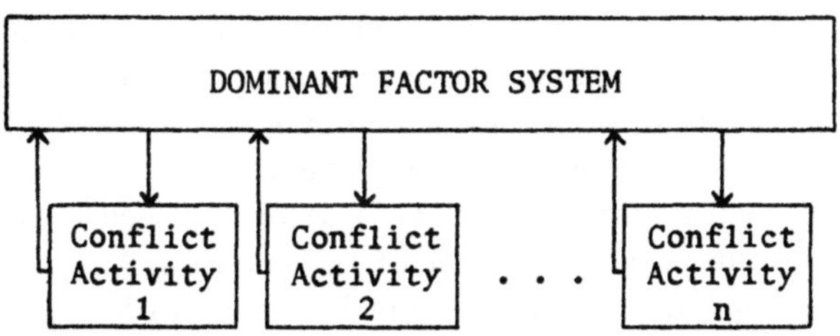

Figure 5.4. Extended Model: Linkages of Dominant Factor System to Conflict Activities

The extended considerations implied by Figure 5.4 are relevant for policy considerations, especially in terms of a perspective that recognizes that the various forms of conflict activity differ substantially with respect to social costs associated with them. In terms of this representation, the view of conflict activities as epiphenomena may be extended to one that recognizes that the events are themselves policy tools employed by and against private groups and governments in pursuit of particular objectives. Difficulties that may be involved in, say, modification of political structure or arresting social disorganization with a focus on the effects of such policy goals on given types of events indicates the need for an orientation that considers effects of all aspects of the overall situation, including the utilization or avoidance of particular forms of conflict activity. For example, the high association of Riots and Revolutionary activity with Purges and Government Crises should affect the utilization of Purges as policy tools. While Purges may provide opportunities for the achievement of apparently greater organizational cohesion and harmony, their interaction with situations of Government Crisis may indicate negative overall effects through Riots and Revolutionary activity.

It is also worth considering that these effects should not be thought of solely in terms of results of so-called dominant factors, but should be investigated also from the point of view of potential direct-interaction effects of conflict activities. Thus, the investigation of the domestic conflict system itself should be extended to an adequate expression of linkages among conflict events. Although it will make more sense to then consider linkages to dominant factors, a truly adequate explication of these relationships would require an overall definition of an object system that includes characteristics related to social context.

The rest of this chapter investigates the domestic conflict system only as defined and attempts to derive a structural or organizational description that best explains phenomena represented in the behavior system. A derivation of this structural description would represent—in the sense described in Chapter 1—system description at a higher epistemological level. That is, while the determined behavior system gives an idea of interaction in the sense of which appearances of which attributes (conflict activities) co-occur and with what frequency, a structural description illustrates the organization among the overall events. This requires a description of the more complex overall events, or the overall system, in terms of simpler events or subsystems such that some property of the overall system is adequately preserved. We observe that such structural description has been a primary goal of most of the empirical factor analytic work referred to earlier and illustrated in Figure 5.1. It is worth observing also that none of the factor structures determined from this previous work gives an adequate expectation of the high-frequency anomalous state that we have discussed and also that none of these past structure determinations has integrated any sense of dynamics into the determination of structure.

In terms of the utilization of GSPS, this latter aspect is easily accomplished through an extended behavior determination—that is, one that includes in the state of the system past as well as current values of the attributes that define the system. The mechanism for accomplishing this is described in Klir (1975) and in Cavallo (1979) and is not repeated here. Cavallo (1979) also includes some results of processing of the data used here from this perspective. Since our current objective with respect to this description involving *memory* is its utilization in evaluation and determination of structure systems, we do not deal here with interpretation of these results. For an adequate study of dynamics of conflict phenomena involving memory at the behavior level, including at least the attributes of guerilla war and coups d'etat, is apparently necessary to allow the comparison and testing of certain hypothetical statements in past literature. For example, Huntington (1970) maintains that guerrilla war represents a developmental stage succeeding revolution in the Western experience but is a precursor of revolution in the experience of Asia. Obviously, a redefinition of the object system and an extension of our data base to include modeling of these attributes could be easily accomplished.

STRUCTURE SYSTEMS

In considering the structure systems that we describe in the following paragraphs, we would like to emphasize again the overall process character of modeling activity. That is, although individual systems problems as formulated have "solutions," it is the integration of these solutions in an interpretively accessible manner that constitutes the primary feature of GSPS. This aspect motivates a related interpretation at each of the problem-definition levels—that is, at each level in the hierarchy—and one that is especially useful for the consideration of structure systems that follows. In terms of the overall modeling activity, we have described an interaction between GSPS—basically defined in terms of general systems concepts in such a manner as to access and make use of formal and abstract constructions—and linguistic and ideational constructs related much more specifically to study of a particular content—that of domestic conflict. The interpretation of modeling gives process aspects the major emphasis.

This characterization is also a useful one from which to consider problems related to specific levels—for example, structure determination. In this view the emphasis on process motivates an interpretation of structure: instead of looking for *the* structure of the system, we recognize that "structure" in a broader sense represents a description resulting from the relation between investigator and the phenomena under investigation. Further, this relation should be able to express itself at all levels and portions of an investigation. Thus, the structure-determination process should be considered as much an elicitation of structural tendencies as a determination of a "best" structure system. Structure systems (refinements of overall systems or description of a whole in terms of subsystems and their interrelations) in general involve loss of information. The most adequate description will always be the full object of investigation and full set of relations on the attributes defining the object. Important questions relate to how completely various refinements capture the overall system, what significance each of a set of refinement stages has in terms of the interpreted specific system it represents, and what significance may accrue to the fact that one refinement of a particular type more adequately expresses the data (as empirically determined) of a whole system than does another. We thus characterize structures as acceptable or meaningful depending on the extent to which interaction among variables that are grouped together in a subsystem is greater than the interaction between variables in different subsystems.

Before illustrating the use of the process we are referring to on this specific system, we first mention some aspects of the process. The motivation for our view of structure determination stems from a recognition that it is in general not valid to assume that an overall system of interest can be adequately described by even perfect representation of parts of systems. Simple and clear illustrations are given, for example, in Cavallo (1979) showing how systems defined with three variables can exhibit significant interaction while each pair of variables exhibits total independence of each of the pairs of associated variables. This theme is also dealt with in a number of papers by Broekstra (e.g., 1976, 1978). It is thus reasonable to expect that techniques based on pairwise correlations will often not detect meaningful interaction that exists. As an alternative to such approaches, the obvious solution is to start with the overall system and generate and evaluate various refinements in terms of this system. All the relevant issues are described in Cavallo (1979). Cavallo and Klir (1979, 1981) contain a complete description of the algorithms whose implementation is a problem-solving tool easily used in this and other investigations. (See also Chapter 1 of this volume.)

If we consider the structure-determination process as a problem within GSPS, the formulation is in terms of a given (initial) specific behavior system representing interaction among all the attributes. The terminal system type is that of structure system, and the requirement types specify a listing of various refinements along with the evaluation of each refinement in terms of specified criteria. The criteria utilized here are (1) the divergence between assumed probability distributions associated with overall states that would result from a hypothesized structure and those that are given as the empirically determined distribution and (2) the added uncertainty that accrues to prediction of the next state of the overall system if a given refinement is hypothesized. These criteria are developed and described in detail in Klir (1976), Klir and Uyttenhove (1976), and Cavallo

(1979). We observe that criterion (1) captures the sense of the extent to which a given structure adequately represents the overall facts or the extent to which the overall situation is reconstructable from partial information (see Cavallo and Klir, 1979, 1981); criterion (2) expresses the extent to which what is likely to occur can be deduced without looking at information regarding all of the interactions – criteria that relate to the discussion following the model in Figure 5.4.

In Cavallo (1979), a sequence of structures determined from this conflict data is presented in terms of the overall refinement process. Three of those structures are reproduced in Figure 5.5. (Where it does not make a difference, attribute indicators are labeled with their symbols as given earlier.)

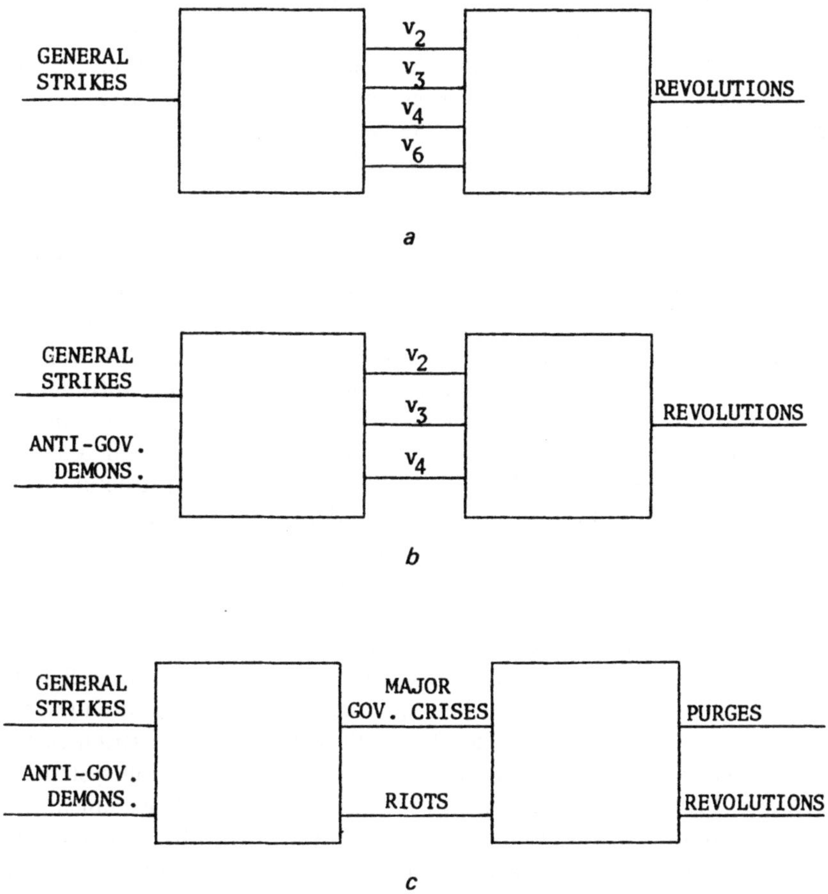

Figure 5.5. Best Structures at Various Levels of Refinement

Figure 5.5*a* is a graphic representation, in the form of Figure 5.1, of the structure from among the class of those that represent the smallest amount of refinement – the dissociation of direct relation between only two variables – from which least information about the overall system is lost. We may interpret this figure to indicate that, beginning with consideration of the conflict system as a whole, the most compelling tendency toward subsystem formation is that which distinguishes Strike activity from that of Revolutions. This is somewhat natural and validates a reasonable expectation that the most distinguishing characteristic of conflict activity may stem from whether the major emphasis is on the overthrow or replacement of the existing political regime. As obvious as this may appear, however, such a determination would not be possible to make from the data and approach used by Rummel (1965), where, for the years of that study, the correlation between Strike and Revolutionary activity was not particularly low. Figure 5.5*b* gives the most acceptable from among the next-least-refined structures and supports the interpretation given with respect to Figure 5.5*a*.

Although a number of other structure systems are generated by the process and may be significant in terms of the overall investigation, the only other one that is considered here is the one given in 5.5*c*. As more interrelations are removed from the overall system, the use of criteria (1) and (2) gives evaluations that get progressively worse. That is, as the form of the structural descriptions gets simpler, more significant information associated with the overall system is lost and the worse is the approximation to the complete system – the worst case being that in which all attributes are considered independently (as is implicitly assumed in the situation represented in Figure 5.3). Thus, trade-offs must be made between approximation and complexity in determination of structure, and they have to be made in terms of the interpretation of the system and of the purposes of the investigation. The purposes here are to derive a structure system that is similar to those utilized in previous work and that facilitates comparison of the approach used here with that work and, more directly, to give a structural "explanation" of the anomaly discovered at the level of behavior. In Cavallo (1979) arguments are given for the acceptance of the structure represented in Figure 5.5*c* as a reasonable compromise between approximation and complexity.

Figure 5.5*c* represents the best refinement that separates the overall system into two interacting subsystems, where the interaction or information transmission between the two subsystems is effected through two of the attributes. Such a description provides a basis from which to consider the overall system. The structure in Figure 5.5*c* associates General Strikes with Anti-Government Demonstrations. Both originate from nongovernment sources and basically constitute efforts to communicate with and establish claims on the government through nonviolent and, for the most part, lawful means. They also reflect the chosen strategy of highly organized groups. General Strikes constitute efforts to communicate dissatisfaction by withdrawing services necessary to the economy as well as basic governmental functions. Anti-Government Demonstrations tend to be more highly articulated and to focus efforts to communicate viewpoints, but without imposing economic costs as in the case of General Strikes.

The structure of Figure 5.5*c* presents these forms of nongovernment-initiated conflict as most remote from Purges and Revolutionary activity. Both are also conceptually remote since they are based on the assumption that the regime in power is open to change through persuasion, and they do not in general constitute a direct assault on the ability of the regime to remain in power. Revolutionary and Purge activities, in contrast, are based on the assumption of mutually irreconcilable differences that can be rectified only by elimination of the opposition. As attributes that couple these two subsystems, Riots and Major Government Crises are seen to occupy a linking position between activities with the two described tendencies. They represent in one case severe expressions of discontent (Riots) and in the other the impact of events on government coping with its environment. These conflict events link the most distant extremes to describe the structure of disturbances to the system. An important indication of the validity of the structure given in Figure 5.5*c* stems from observing that the structure as described represents and gives an explanation for the high occurrence of the Riots/Revolutions/Purges/Crises state. None of the previously determined factor-analytic structures gives an indication of the importance of this combination.

We may also observe that from the specific systems point of view, the represented structure also provides a point of departure for consideration of one of the basic issues associated with domestic conflict—that dealing with elite circulation. Inability to interpret and accept discontent of sufficient severity to precipitate Major Government Crises may reflect a failure of the ruling elite that could lead to replacement through Purges and Revolution. This interpretation of the data, behavior, and structure systems associated with conflict phenomena provides empirical support and a framework for study of elite behavior along the lines of observations originally made by Pareto (1963), who divides established elites into groups that are willing to defend their position through use of force and those who do not exhibit this orientation. Similarly, counterelites may be divided into two groups distinguished by their willingness to use force to replace existing elites, a perception also supported by the derived structure system.

Although all of these indications require further elaboration and study, it is not unreasonable to expect that the complex and dynamic nature of the specific systems involved preclude the possibility that a closed theoretical statement could be adequate to describe the situation totally. If this perspective is accepted (cf. Hayek, 1974), the importance of a commonly accepted general methodological approach is a crucial component in the meaningful investigation of these systems. The considerations of this paper indicate that an overall systems investigative approach as described here—that is, that results of information processing that involves the integrated use of general and comprehensive techniques—represents a far more adequate basis for specific systems-theoretical constructions and for social system policy analysis than could otherwise be achieved.

REFERENCES

Azar, E.E., and J. Ben-Dak, eds. 1975. *Theory and Practice of Events Research.* New York: Gordon & Breach.

Banks, A.S. 1972. "Patterns of Domestic Conflict: 1919-39 and 1946-66." *Journal of Conflict Resolution* 16:41-50.

Barto, A., and P. Davis. 1978. "Dyadic Consensus-Competition Models." Internal working paper, State University of New York at Binghamton, School of Advanced Technology.

Brinton, C.C. 1956. *The Anatomy of Revolution.* New York: Vintage Books.

Broekstra, G. 1976. "Constraint Analysis and Structure Identification." *Annals of Systems Research* 5:67-80.

——. 1978. "On the Representation and Identification of Structure Systems." *International Journal of Systems Science* 9(11):1271-93.

Cavallo, R.E. 1979. *The Role of Systems Methodology in Social Science Research.* Boston: Martinus Nijhoff.

Cavallo, R.E., and G.J. Klir. 1979. "Reconstructability Analysis of Multi-Dimensional Relations: A Theoretical Basis for Computer-Aided Determination of Acceptable Systems Models." *International Journal of General Systems* 5(3):143-71.

——. 1981. "Reconstructability Analysis: Evaluation of Reconstruction Hypotheses." *International Journal of General Systems* 7(1):7-32.

Davies, J.C. 1963. *Human Nature in Politics.* New York: Wiley.

Deutsch, K.W. 1963. *The Nerves of Government.* New York: Free Press.

Downes, B. 1968. "Social and Political Characteristics of Riot Cities: A Comparative Study." *Social Science Quarterly* 49:504-20.

Edwards, L. 1970. *The Natural History of Revolution.* Chicago: University of Chicago Press.

Feierabend, I., and R. Feierabend. 1966. "Aggressive Behaviors within Politics, 1948-1962: A Cross National Study." *Journal of Conflict Resolution* 10.

Firestone, J., and D. McCormick. 1972. "An Exploration in Systems Analysis of Domestic Conflict." *General Systems* 17:79-119.

Gurr, T.R., 1968. "A Causal Model of Civil Strife: A Comparative Analysis Using New Indices." *American Political Science Review* 67.

Gurr, T.R., and V.F. Bishop. 1976. "Violent Nations, and Others." *Journal of Conflict Resolution* 20(1):79-110.

Hayek, F.A. 1974. "The Pretence of Knowledge." (Nobel Memorial Prize

Lecture.) In *Full Employment at Any Price?* London: The Institute of Economic Affairs.

Hibbs, D.A. 1973. *Mass Political Violence: A Cross National Causal Analysis.* New York: Wiley.

Huntington, S.P. 1970. *Political Order in Changing Societies.* New Haven, Conn.: Yale University Press.

Klir, G.J. 1975. "On the Representation of Activity Arrays." *International Journal of General Systems* 2(3):149–68.

——. 1976. "Identification of Generative Structures in Empirical Data." *International Journal of General Systems* 3(2):89–104.

Klir, G., and H. Uyttenhove. 1976. "Computerized Methodology for Structure Modelling." *Annals of Systems Research* 5:29–66.

Kornhauser, W. 1959. *The Politics of Mass Society.* Glencoe, Ill.: Free Press.

Lupsha, P. 1969. "On Theories of Urban Violence." *Urban Affairs Quarterly* (March):273–96.

Pareto, V. 1963. *The Mind and Society: A Treatise on General Sociology.* New York: Dover.

Pettee, G. 1938. *The Process of Revolution.* New York: Harper & Row.

Rogers, G. 1977. *Computer Simulations of Cellular Movements.* Ph.D. dissertation, State University of New York at Binghamton, School of Advanced Technology.

Rummel, R.J. 1963. "Dimensions of Conflict Behavior within and between Nations." *General Systems* 8:1–50.

——. 1965. "A Field Theory of Social Actions with Application to Conflict within Nations." *General Systems Yearbook* 10:183–211.

——. 1970. *Applied Factor Analysis.* Evanston, Ill.: Northwestern University Press.

Singer, J.D. 1976. "Rejoinder to the Critiques of the Correlates of War Project." In F.W. Hoole and D.A. Zinnes, eds. *Quantitative International Politics: An Appraisal.* New York: Praeger.

Spiegel, J.P. 1971. "Theories of Violence: An Integrated Approach." *International Journal of Group Tensions* 1:77–90.

Tanter, R. 1966. "Dimensions of Conflict Behavior within and between Nations, 1958–1960." *Journal of Conflict Resolution* 10:41–64.

Terrell, L.M. 1971. "Societal Stress, Political Instability, and Levels of Military Effort." *Journal of Conflict Resolution* 15:237–54.

6 THEORY OF MEASUREMENT OF IMPACTS AND INTERACTIONS IN SYSTEMS

Thomas L. Saaty

EIGENVALUES, PRIORITIES, HIERARCHIES, AND NETWORKS

What is being frequently tested in society today is not our engineering skill or our ability to bring technology to bear on life's problems, but our sense of priorities and our capacity to make trade-offs between various factors that have an impact on life. The trade-offs cannot all be measured in dollars. Rather, the priorities themselves provide the scales in each of the areas, and where dollars are concerned they should indeed be measured. What we have been looking for is a way of determining priorities that does not preclude our previous approach to economic cost-benefit analysis but generalizes it to include social, political, and environmental factors as well. We will first derive priorities for the elements in a system whose components are hierarchically arranged.

To do this we develop a theory to arrive at ratio scales by solving the largest eigenvalue problems for pairwise comparison matrices. The effects of initial judgment consistency on the final priorities are of particular concern. We illustrate this approach to measurement with an example and then develop the principle of composition of priorities to be applied to hierarchical structures. Finally, we generalize the composition principle to systems, allowing for the possibility of feedback and cycling. From this analysis we obtain priorities for the impacts of the components of the system on the physical, social, political, environmental, and other important properties to which the system must conform in pursuit of its objectives.

The numbers we get compare very well with measurements for which a scale has been established in real life. For example, we have used our method to compare the weights of objects and have found that the largest eigenvalue solution produces an accurate estimate of their relative weights. For a more sophisticated application, see the GNP example included here. This method of establishing priorities has been applied to national transport planning, energy allocation to industries, corporate and military planning, and a very wide variety of individual problems ranging from job selection to school and housing choice and even to psychotherapy.

RATIO SCALE ESTIMATION

Suppose that we have a set of n people, $P_1, \ldots, P_n$. We want to estimate their relative heights by pairwise comparisons. For the moment let us assume that their heights are known to be respectively, $w_1, \ldots, w_n$. Then pairwise comparison yields the matrix of ratios $A = (w_i/w_j)$, where the ith person represented by a row is compared with the jth person represented by a column. We note that we can recover the scale $w = (w_1, \ldots, w_n)$ by first forming the product Aw, which simplifies to nw, and then solving the problem $Aw = nw$. The set of homogeneous equations has a nontrivial solution if and only if n is an eigenvalue of A. Now A has unit rank since every row is a constant multiple of every other row, say, for example, the first row. The ith row is obtained by multiplying the first row by w_i/w_1. Thus, all the eigenvalues $\lambda_1, \lambda_2, \ldots, \lambda_n$ of A are zero except for one. Also, $\Sigma_{i=1}^{n} \lambda_i = \text{trace } A = n$. Hence n is the largest eigenvalue of A, and the problem has a nontrivial solution. Now A satisfies the property $a_{ij}a_{jk} = a_{ik}$ (i.e., all its entries may be generated from a spanning cycle of n elements). It is said to be consistent.

Let us now assume that the ratios w_i/w_j are not known, but that we have estimates of them. Let us also assume that these estimates are provided by an expert, and hence they are close to their true values. In that case we have a small perturbation of the a_{ij}, which (from matrix theory) we know is accompanied by a small perturbation of the λ's. Now A no longer need be consistent. Note that our original A has the property $a_{ji} = 1/a_{ij}$ and is called a reciprocal matrix. In other words, if one person is a times taller than another, then the latter is $1/a$ times as tall as the first—a reasonable assumption. Thus in the new matrix, we only estimate a_{ij} and simply enter $a_{ji} = 1/a_{ij}$. We have shown that a reciprocal matrix A is consistent if, and only if, $\lambda_{max} \equiv \lambda_1 = n$. We have also shown that for such a matrix $\lambda_{max} \geqslant n$.

In case A has linear elementary divisors, we may assume the existence of a complete set of eigenvectors $w_1, \ldots, w_n$, which are unique only if all eigenvalues $\lambda_1, \ldots, \lambda_n$ are simple. (We know only that $\lambda_1 = \lambda_{max}$ is simple.) Thus $w_1, \ldots, w_n$ form a complete orthonormal basis in terms of which any vector can be expressed as a linear combination.

In particular:

$$e = a_1 w_1 + \cdots + a_n w_n \qquad e^T = (1, 1, \ldots, 1);$$

$$Ae = a_1 \lambda_1 w_1 + \cdots + a_n \lambda_n w_n;$$

$$A^k e = a_1 \lambda_1^k w_1 + \cdots + a_n \lambda_n^k w_n.$$

Since $\lambda_1 \geqslant |\lambda_i|$, for all i we have

$$A^k e = \lambda_1^k \left(a_1 w_1 + \cdots + a_n \left(\frac{\lambda_n}{\lambda_1} \right)^k w_n \right)$$

and

$$\lim_{k \to \infty} \lambda_1^{-k} A^k e = a_1 w_1.$$

This can be shown to hold with more general assumptions on A.

Thus the eigenvector w_1 corresponding to λ_{max} is the normalized limiting vector of the row sums of powers of A. This is indeed the way to compute w_1 in practice by raising A to powers A^{2k} and stopping when the error between two successive powers does not exceed a prescribed value.

The departure of λ_{max} from n (i.e., $\lambda_{max} - n$ [actually we use $(\lambda_{max} - n)/(n - 1) = -(1/n - 1)\Sigma_{i=2}^{n} \lambda_i$]) serves as a measure of departure from consistency that can occur due to violations of $a_{ij}a_{jk} = a_{ik}$ or simply to intransitivity in judgments. An easy way to observe the relation between departure from consistency and $\lambda_{max} - n$ is to form

$$a_{ik}a_{kj} - a_{ij}$$

and to sum over k, obtaining

$$\sum_{k=1}^{n} a_{ik}a_{kj} - na_{ij}.$$

This matrix is simply $A^2 - nA$, which, according to Sylvester's decomposition theorem for simple eigenvalues, is equal to

$$\sum_{i=1}^{n} (\lambda_i^2 - n\lambda_i)Z(\lambda_i),$$

where $Z(\lambda_i)$ are orthogonal idempotent matrices, thus involving $\lambda_{max}^2 - n\lambda_{max}$. More simply, operating on the eigenvector w of A, corresponding to λ_{max}, yields $(A^2 - nA)w = \lambda_{max}(\lambda_{max} - n)w$. From this it is clear that when A is consistent (for a consistent matrix $A^k = n^{k-2}A$), $\lambda_{max} = n$, and the departure from consistency is reflected in $\lambda_{max}(\lambda_{max} - n)$ or simply in $\lambda_{max} - n = -\Sigma_{i=2}^{n} \lambda_i$. Its average value with respect to λ_i, $i = 2, \ldots, n$ is given by $(\lambda_{max} - n)/(n - 1)$.

REMARK. It may be interesting to compare the maximum eigenvalue formulation with the least squares formulation, which is concerned with minimizing

$$\text{trace}\,(A - X)(A - X)^T = \sum_{i,j=1}^{n} \left(a_{ij} - \frac{x_i}{x_j}\right)^2 = \sum_{i=1}^{n} \alpha_i,$$

where $X = (x_i/x_j)$ and α_i, $i = 1, \ldots, n$ are the eigenvalues of $(A - X)(A - X)^T$. The latter matrix is symmetric, and all its eigenvalues are real. With consistency, both methods give the same solution. We have seen that the maximum eigenvalue approximation is obtained by a perturbation argument on $Aw = nw$, the paradigm case. There is no simple expression relating $w = (w_1, \ldots, w_n)$ and $x = (x_1, \ldots, x_n)$, but with near consistency w turns out to be a better approximation of the underlying scale than x, in the root mean square sense. (For further details see Gantmacher, 1960, and Saaty, 1977.)

Note that the eigenvalue approach provides one with a measure of consistency. Thus, even when the estimate is not given by an expert, we may accept or reject the result of solving $Aw = \lambda_{max}w$, depending on the size of $(\lambda_{max} - n)/(n - 1)$. The value of this index is compared with its mean value from a sample of fifty reciprocal matrices of the same order whose entries are obtained at random from a numerical scale to be discussed below. If the ratio is small, the data are accepted. Otherwise, more information would be needed to improve the consistency. Note that consistency is a necessary but not a sufficient condition for modeling validly. To validate the result in practice, one usually needs some real data for comparison purposes and statistical measure of closeness of fit. We have been using the root mean square deviation and the median absolute deviation about the median for this purpose. We have proved that the eigenvector is insensitive to small perturbations in judgments.

THE SCALE

The scale we recommend for use, which has been successfully tested and compared with other scales (Miller, 1956; Saaty, 1977), will now be discussed. The judgments elicited from people are taken qualitatively, and corresponding scale values are assigned to them. In general, we do not expect "cardinal" consistency to hold everywhere in the matrix because people's feelings do not conform to an exact formula. Nor do we expect "ordinal" consistency, since people's judgments may not be transitive. However, to improve consistency in the numerical judgments, whatever value a_{ij} is assigned in comparing the ith activity with the jth one, the reciprocal value is assigned to a_{ji}. Thus we put $a_{ji} = 1/a_{ij}$. Usually, we first record whichever value represents dominance greater than unity.

Since we require that the subject must be aware of all gradations at the same time, and we agree with the psychological experiments (Miller, 1956) that show an individual cannot simultaneously compare more than seven objects (plus or minus two) without being confused, we are led to choose a $p = 7 + 2$. Using a unit difference between successive scale values is all that we allow, and using the fact that $x_1 = 1$ for the identity comparison, it follows that the scale values will range from 1 to 9.

As a preliminary step toward the construction of an intensity scale of importance for activities, we have broken down the importance ranks as shown in Table 6.1. In using this scale the reader should recall that we assume that the individual providing the judgment has knowledge about the relative values of the elements being compared (whose ration is $\geqslant 1$), and that the numerical ratios formed are nearest-integer approximations scaled in such a way that the highest ratio corresponds to 9. We have assumed that an element with weight zero is eliminated from comparison. Reciprocals of all scaled ratios that are $\geqslant 1$ are entered in the transpose positions (not taken as judgments). In passing, we note that the eigenvector solution of the problem remains the same if we multiply the unit entries on the main diagonal by a constant greater than one.

At first glance we would like to have a scale extend as far out as possible. On second thought we discover that to give an idea how large measurement can get, scales must be finite. We also note that one does not measure widely disparate objects by the same yardstick. Short distances on a piece of paper are measured in centimeters, longer distances in a neighborhood in meters, and still larger ones in kilometers and even in light years. When we compare the sizes of atoms with those of stars, we tend to insert between these extremes objects that gradually grow larger and larger so that one can appreciate the transition in the magnitudes of measurement. To make such a transition possible, the objects are divided into groups or clusters whereby the objects put in each group are within

Table 6.1. The Scale and Its Description

Intensity of Importance	*Definition*	*Explanation*
1*	Equal importance	Two activities contribute equally to the objective
3	Weak importance of one over another	Experience and judgment slightly favor one activity over another
5	Essential or strong importance	Experience and judgment strongly favor one activity over another
7	Demonstrated importance	An activity is strongly favored and its dominance demonstrated in practice
9	Absolute importance	The evidence favoring one activity over another is of the highest possible order of affirmation
2, 4, 6, 8	Intermediate values between the two adjacent judgments	When compromise is needed
Reciprocals of above nonzero	If activity i has one of the above nonzero numbers assigned to it when compared with activity j, then j has the reciprocal value when compared with i	
Rationals	Ratios arising from the scale	If consistency were to be forced by obtaining n numerical values to span the matrix

*On occasion in 2×2 problems, we have used $1 + \epsilon$, $0 < \epsilon \leqslant 1/2$ to indicate very slight dominance between two nearly equal activities.

the range of the scale, and the largest object in one group is used as the smallest one in the next larger group. Its scale values in the two groups enable one to continue the measurement from one group to the next and so on. We shall have more to say about clustering below. In practice, one way or another, the numerical judgments will have to be approximations, but how good is the question at which our theory is aimed?

A typical question to ask when filling in the entries in a matrix of comparisons is this: Consider the two properties i on the left side of the matrix and another j on the top; which of the two has the property under discussion more, and how strongly more (using the same values 1 to 9)? This gives us a_{ij}. The reciprocal value is then automatically entered for a_{ji}.

Considerable effort has been concentrated on comparing the scale 1 to 9 with twenty-five other scales suggested to us by a number of people. We took pairwise qualitative judgments described in our scale, including qualities between those mentioned in the table.

Example: The Wealth of Nations through Their World Influence

A number of people have studied the problem of measuring world influence of nations. We have briefly examined this concept within the framework of our model. We assumed that influence is a function of several factors. We considered five such factors (Saaty and Khousa, 1976): (1) human resources, (2) wealth, (3) trade, (4) technology, and (5) military power. Culture and ideology, as well as potential natural resources (such as oil), were not included.

We selected seven countries—United States, Soviet Union, China, France, United Kingdom, Japan, and West Germany—for this analysis because we believed that these nations as a group comprised a dominant class of influential nations. We wished to compare them among themselves as to their overall influence in international relations. We realize that what we have is a very rough estimate, mainly intended to serve as an interesting example of an application of our approach to priorities. We will illustrate the method with respect to only the single factor of wealth.

We now give the matrix in Table 6.2 of pairwise comparisons of the seven countries with respect to wealth. For example, the value 4 in the first row indicates that wealth is between weak and strong importance in favor of the United States over the Soviet Union. The reciprocal of 4 appears in the symmetric position, indicating the inverse relation of relative strength of the wealth of the Soviet Union compared with the United States.

Table 6.2. Wealth Comparison of Nations

	United States	*Soviet Union*	*China*	*France*	*United Kingdom*	*Japan*	*West Germany*
United States	1	4	9	6	6	5	5
Soviet Union	1/4	1	7	5	5	3	4
China	1/9	1/7	1	1/5	1/5	1/7	1/5
France	1/6	1/5	5	1	1	1/3	1/3
United Kingdom	1/6	1/5	5	1	1	1/3	1/3
Japan	1/5	1/3	7	3	3	1	2
West Germany	1/5	1/4	5	3	3	1/2	1

Explanation of Table 6.2

The first row of Table 6.2 gives the pairwise comparison of the wealth of the United States with that of the other nations. For example, it is of equal importance to the United States (hence, the unit entry in the first position), between weak and strong importance when compared with the Soviet Union (hence, the value 4 in the second position), and of absolute importance when compared with China (hence, the value 9 in the third position). We have values between strong and demonstrated importance when compared with France and the United Kingdom (hence, a 6 in the next two positions), and of strong importance when compared with Japan and Germany (hence a 5 in the following two positions). For the entries in the first column, we have the reciprocals of the numbers in the first row, indicating the inverse relations of relative strength of the wealth of the other countries when compared with the United States and so on for the remaining values in the second row and second column, etc.

Note that the comparisons are not consistent. For example, United States compared with Soviet Union equals 4, Soviet Union compared with China equals 7, but United States compared with China equals 9, not 28. Nevertheless, when the requisite computations are performed, we obtain relative weights of 42.9 and 23.1 for the United States and Soviet Union. These weights are in striking agreement with the corresponding GNP's as percentages of the total GNP (see Table 6.3). Thus, despite the apparent arbitrariness of the scale, the irregularities disappear and the numbers occur in good accord with observed data. The largest eigenvalue of the wealth example is 7.61, and the remaining ones are $-.228$; 2×10^{-16}; $-.330 + .588i$; $-.330 - .588i$; $.14 + 2.06i$; $.14 - 2.06i$. Compare the normalized eigenvector column derived by using the matrix of judgments in

Table 6.3. Normalized Wealth Eigenvector

	Normalized Eigenvector	*Actual GNP* (1972)*	*Fraction of GNP Total*
United States	.429	1,167	.413
Soviet Union	.231	635	.225
China	.021	120	.043
France	.053	196	.069
United Kingdom	.053	154	.055
Japan	.119	294	.104
West Germany	.095	257	.091
Total		2,823	

*In billions of dollars.

Table 6.3 with the actual GNP fraction given in the last column. The two are very close in their values. Estimates of the actual GNP of China range from 74 billion to 128 billion. The value for China is more than it is for Japan in that our estimate is half the (admittedly uncertain) GNP value. Japan's value is a third over the true value. Cluster analysis can be used to show that China probably should not be in the group.

The Wealth Example as a Cluster

The comparison of the wealth of seven nations was made by means of clustering the nations into three groups: A = (United States), B = (Soviet Union), and C = (United Kingdom, France, Japan, West Germany). The clusters were first compared, yielding the following matrix:

	A	B	C	*Eigenvector*
A	1	2	1	.4
B	1/2	1	1/2	.2
C	1	2	1	.4

$\lambda_{max} = 3$

The elements of C were compared among themselves in the following matrix:

	United Kingdom	*France*	*Japan*	*West Germany*	*Eigenvector*
United Kingdom	1	1	1/3	1/2	.14
France	1	1	1/3	1/2	.14
Japan	3	3	1	2	.45
West Germany	2	2	1/2	1	.26

The estimated relative wealth obtained this way is given by:

United States	*Soviet Union*	*United Kingdom*	*France*	*Japan*	*West Germany*
.4	.2	.056	.056	.18	.10

FORMAL HIERARCHIES

We use the notation $x^- = \{y | x \text{ covers } y\}$ and $x^+ = \{y | y \text{ covers } x\}$ for any element x in an ordered set. (See also Saaty, 1977.)

DEFINITION. Let H be a finite partially ordered set with largest element b. H is a *hierarchy* if it satisfies the following conditions:

1. There is a partition of H into sets $L_k, k = 1, \ldots, h$, where $L_1 = \{b\}$.
2. $x \in L_k$ implies $x^- \subset L_{k+1}$ $\quad k = 1, \ldots, h - 1$.
3. $x \in L_k$ implies $x^+ \subset L_{k-1}$ $\quad k = 2, \ldots, h$.

For each $x \in H$, there is a suitable weighting function (whose nature depends on the phenomenon being hierarchically structured):

$$w_x : x^- \rightarrow [0,1] \text{ such that } \sum_{y \in x^-} w_x(y) = 1.$$

The sets L_i are the *levels* of the hierarchy, and the function w_x is the *priority function* of the elements in one level with respect to the objective x. We observe that even if $x^- \neq L_k$ (for some level L_k), w_x may be defined for all of L_k by setting it equal to zero for all elements in L_k not in x^-.

The weighting function, we believe, is a significant contribution toward the application of hierarchy theory.

DEFINITION. A hierarchy is *complete* if, for all $x \in L_k$, $x^+ = L_{k-1}$, for $k = 2, \ldots, h$.

We can state the central question:

Basic problem. Given any element $x \in L_\alpha$, and subset $S \subset L_\beta$ $(\alpha < \beta)$, how do we define a function $w_{x,S} : S \to [0,1]$ that reflects the properties of the priority functions w_y on the levels L_k, $k = \alpha, \ldots, \beta - 1$? Specifically, what is the function $w_{b,L_h} : L_h \to [0,1]$? The problem can be paraphrased in less technical terms: Given a social (or economic) system with a major objective, b, and the set L_h of basic activities, such that the system can be modeled as a hierarchy with largest element b and lowest level L_h, what are the priorities of the elements of L_h with respect to b?

From the standpoint of optimization, to allocate a resource among the elements, any interdependence must also be considered. Analytically, interdependence may take the form of input-output relations, such as the interflow of products between industries. A high-priority industry may depend on flow of material from a low-priority industry. In an optimization framework the priority of the elements enables one to define the objective function to be maximized, and other hierarchies supply information regarding constraints (e.g., input-output relations).

We shall now present our method to solve the basic problem. Assume that $Y = \{y_1, \ldots, y_{m_k}\} \in L_k$ and that $X = \{x_1, \ldots, x_{m_{k+1}}\} \in L_{k+1}$. (Observe that, according to the remark following the definition of hierarchy, we may assume that $X = L_{k+1}$.) Also assume that there is an element $z \in L_{k-1}$, such that $y \subset z^-$. We then consider the priority functions

$$w_z : Y \to [0,1] \text{ and } w_{y_j} : X \to [0,1] \qquad j = 1, \ldots, n_k.$$

We construct the "priority function of the elements in X with respect to z," denoted w, $w : X \to [0,1]$, by

$$w(x_i) = \sum_{j=1}^{n_k} w_{y_j}(x_i) w_z(y_j), \qquad i = 1, \ldots, n_{k+1}.$$

It is obvious that this is no more than the process of weighting the influence of the element y_j on the priority of x_i by multiplying it with the importance of y_i with respect to z.

The algorithms involved will be simplified if one combines the $w_{y_j}(x_i)$ into a matrix B by setting $b_{ij} = w_{y_j}(x_i)$. If we further set $w_i = w(x_i)$ and $w_j' = w_z(y_j)$, then the above formula becomes

$$w_i = \sum_{j=1}^{n_k} b_{ij} w'_j \qquad i = 1, \ldots, n_{k+1}.$$

Thus, we may speak of the *priority vector, w,* and, indeed, of the *priority matrix* B of the $(k + 1)$st level; this gives the final formulation

$$w = Bw'.$$

The following theorem is easy to prove:

THEOREM. *Let H be a complete hierarchy with largest element b and h levels. Let B_k be the priority matrix of the* kth *level, $k = 1, \ldots, h$. If w' is the priority vector of the* pth *level with respect to some element z in the $(p - 1)$*st *level, then the priority vector w of the* qth *level $(p < q)$ with respect to z is given by*

$$w = B_q B_{q-1} \cdots B_{p+1} w'.$$

Thus, the priority vector of the lowest level with respect to the element b is given by

$$w = B_h B_{h-1} \cdots B_2 b_1.$$

If L_1 has a single element, $b_1 = 1$. Otherwise b_1 is a prescribed vector.

The following observation holds for a complete hierarchy, but it is also useful in general. The priority of an element in a level is the sum of its priorities in each of the comparison subsets to which it belongs; each is weighted by the fraction of elements of the level that belong to that subset and by the priority of that subset. The resulting set of priorities of the elements in the level is then normalized by dividing by its sum. The priority of a subset in a level is equal to the priority of the dominating element in the next level.

We have a large number of examples for the application of our method. The following example has proved to be popular in public presentations due to its simplicity and the fact that it compresses all the elements of the method for a three-level hierarchy.

Example: School Selection

Three high schools—A, B, and C—were analyzed from the standpoint of a candidate according to their desirability. The following six characteristics were selected for the comparison: (1) learning, (2) friends, (3) school life, (4) vocational training, (5) college preparation, and (6) music classes. The pairwise judgment matrices

Table 6.4. Comparison of Characteristics with Overall School Satisfaction

	Learning	*Friends*	*School Life*	*Vocational Training*	*College Preparation*	*Music Classes*
Learning	1	4	3	1	3	4
Friends	1/4	1	7	3	1/5	1
School life	1/3	1/7	1	1/5	1/5	1/6
Vocational training	1	1/3	5	1	1	3
College preparation	1/3	5	5	1	1	3
Music classes	1/4	1	6	1/3	1/3	1

are presented in Table 6.4, and the three schools are compared in the following matrices:

Learning

	A	B	C
A	1	1/3	1/2
B	3	1	3
C	2	1/3	1

Friends

	A	B	C
A	1	1	1
B	1	1	1
C	1	1	1

School Life

	A	B	C
A	1	5	1
B	1/5	1	1/5
C	1	5	1

Vocational Training

	A	B	C
A	1	9	7
B	1/9	1	1/5
C	1/7	5	1

College Preparation

	A	B	C
A	1	1/2	1
B	2	1	2
C	1	1/2	1

Music Classes

	A	B	C
A	1	6	4
B	1/6	1	1/3
C	1/4	3	1

The eigenvector of the first matrix is given by:

$$(.32, .14, .04, .13, .24, .14),$$

and its corresponding eigenvalue is $\lambda = 7.49$, which is far from the consistent value 6. No revision of the matrix was made. Normally, such inconsistency would indicate that we should reconsider the arrangements.

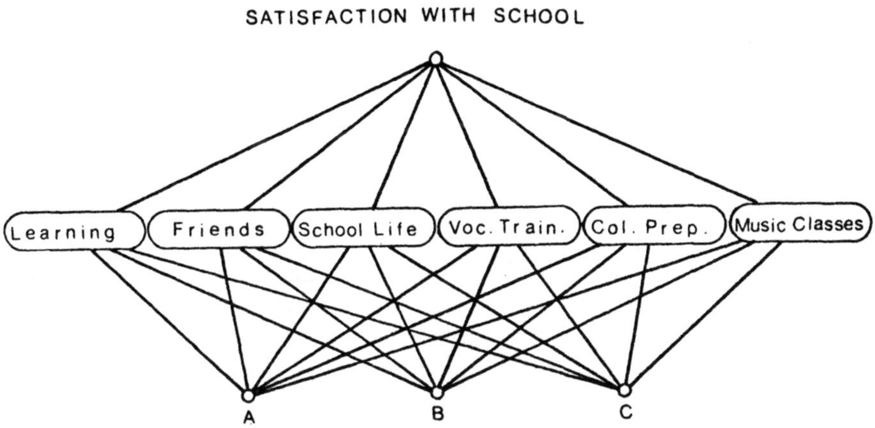

Figure 6.1. School Satisfaction Hierarchy

The eigenvalues and eigenvectors of the other six matrices are:

$\lambda = 3.05$	$\lambda = 3$	$\lambda = 3$	$\lambda = 3.21$	$\lambda = 3.00$	$\lambda = 3.05$
Learning	*Friends*	*School Life*	*Vocational Training*	*College Preparation*	*Music Classes*
.16	.33	.45	.77	.25	.69
.59	.33	.09	.05	.50	.09
.25	.33	.46	.17	.25	.22

To obtain the overall ranking of the schools, we multiply the last matrix on the right by the transpose of the vector of weights of the characteristics. This yields: A = .37; B = .38; C = .25. The individual went to school A because it had almost the same rank as school B, yet school B was a private school charging close to $1,600 a year and school A was free. This is an example where we were able to bring in a lower-priority item (e.g., the cost of the school) to add to the argument that A is favored by the candidate. The actual hierarchy is shown in Figure 6.1.

NETWORKS: SYSTEMS AND FEEDBACK

A system is a set of interacting components. Schematically it is a directed graph whose nodes represent the components and whose arcs indicate the direction of impact of its initial node component on its terminal node component. The impacts are measured as above, with the elements of an initial node component of an arc prioritized according to their impact on each of the elements of a terminal node component. The set of eigenvectors defines a matrix A_{ij} of the impact of the ith component on the jth one, with each column giving the impact on one of the elements of the jth component.

The problem here is to derive a meaningful measure for the impact of components on the entire system. Note that two components may be incident with two oppositely oriented arcs, giving rise to a cycle. Cycles occur in a feedback process between components. A component may have an impact on itself just as it can on another component. This phenomenon has a useful real-life counterpart in input-output analysis in economics. In electric circuits the field generated in a component can affect its own behavior, as would heat from its elements. An example of a system with feedback represented by a directed graph whose nodes are the components is illustrated in Figure 6.2.

We may represent the interactions between components in terms of their impact matrices. The totality of impact matrices A_{ij} defines a super matrix Λ. For example, for a hierarchy H, the matrix Λ has the subdiagonal form, with v_i denoting the nodes of the graph corresponding to the components of the system with their elements arranged to correspond to the columns of the A_{ij}:

$$\Lambda_H = \begin{array}{c} \\ v_1 \\ v_2 \\ v_3 \\ \cdot \\ \cdot \\ \cdot \\ v_n \end{array} \begin{pmatrix} v_1 & v_2 & v_3 & v_4 & \cdots & v_{n-1} & v_n \\ 0 & 0 & 0 & 0 & \cdots & 0 & 0 \\ A_{21} & 0 & 0 & 0 & \cdots & 0 & 0 \\ 0 & A_{32} & 0 & 0 & \cdots & 0 & 0 \\ \cdot & & & & & & \cdot \\ \cdot & & & & & & \cdot \\ \cdot & & & & & & \cdot \\ 0 & 0 & 0 & 0 & \cdots & A_{n,n-1} & 0 \end{pmatrix}.$$

Determining the existence of a limiting or steady-state matrix would be desirable for a system with feedback. Now if a matrix has no block of zeros, it can be multiplied by $1/n$ to become a stochastic matrix, since the sum of the columns of each block matrix is unity. Otherwise, columns are weighted by the number of their nonzero blocks. It is known that a limiting solution exists for such a matrix if it is primitive. An irreducible matrix C is primitive if there is an integer $p \geqslant 1$ such that $A^p > 0$. A matrix is irreducible if its graph is strongly connected (i.e., every arc belongs to a cycle). Thus, for a steady-state solution one must study limiting priorities, much as is done in the case of Markov chains. For reducible systems, such as a hierarchy, one is mainly interested in finite powers of the matrix Λ. For example, the hierarchical composition principle for a complete hierarchy gives the only nonzero entry $A_{n,n-1}A_{n-1,n-2}, \ldots,$

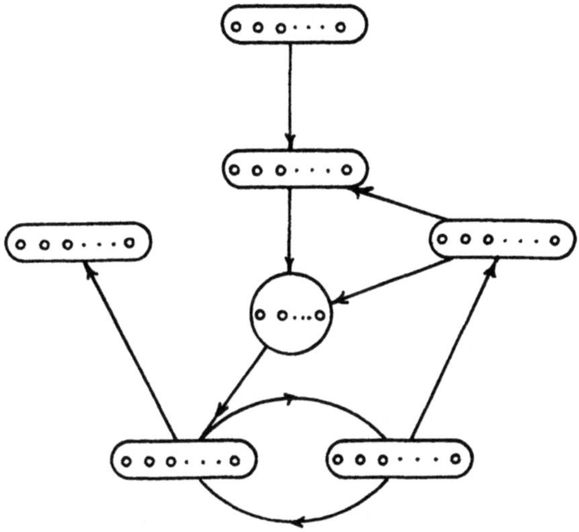

Figure 6.2. System with Feedback

$A_{32}A_{21}$ for the composite eigenvector obtained by raising Λ_H to the $(n - 1)$st power. Applications have been made to systems with feedback in conflict analysis in which a set of actors, by pursuing its objectives and policies, determines a set of outcomes. The powers of the actors are determined by their influence to produce the outcomes. The outcomes in turn have impact priorities on each of the actors, the policies on each of the objectives, and the objectives on each of the actors, thus completing the cycle. Here one is usually interested in finding the highest priority or most likely outcome to emerge as a resultant of the power applied by the various actors.

The usual approach to planning is to project *forward* to what seems *feasible.* Sometimes people have concentrated on proposing a *desired* future and have worked *backward* to determine the means to bring about such a future. For greater effectiveness in designing a plan, these two processes should be combined in a single forward-backward planning process. It has two boundary constraints: one fixed at the present (by the actors and available resources) and one at the future (by the desired outcomes). This two-point boundary process can be treated as a feedback process and has proved very useful in application.

We have made extensive application of the measurement theory to planning and conflict resolution by assisting corporations to devise new policies from the backward process and to test their effectiveness in the forward process (Alexander and Saaty, 1977). Recent applications (with implementation in some cases) have been made to the political problem of Northern Ireland and to numerous examples of terrorism (e.g., the 1972 Munich massacre, the case of the Hanafi

Moslems, Washington, D.C., March 1977, and the Tupamaro kidnapping and killing of the American citizen Daniel De Mitrione in Uruguay, 1970). The upshot of this analysis in recent research for the Arms Control and Disarmament Agency is to establish methods for tackling terrorism problems by prevention or by negotiation after they occur.

REFERENCES

Alexander, J.M., and T.L. Saaty. 1977. "The Forward and Backward Processes of Conflict Analysis." *Behavioral Science* 22(2):87–98.

Gantmacher, F.R. 1960. *The Theory of Matrices,* vol. 2. New York: Chelsea.

Miller, G.A. 1956. "The Magical Number Seven, Plus or Minus Two: Some Limits on our Capacity for Processing Information." *Psychological Review* 63:81–97.

Saaty, T.L. 1977. "A Scaling Method for Priorities and Hierarchical Structures." *Journal of Mathematical Psychology* (June).

Saaty, T.L., and M.W. Khouja. 1976. "A Measure of World Influence." *Journal of Peace Science* 2(1):31–48.

III SYSTEMS-BASED TOOLS FOR SOCIAL SCIENCE RESEARCH

Each of the three papers in this section describes major impacts that a systems orientation has had on the development—from within the social and behavioral sciences—of fundamental tools for the investigation of social systems. Chapter 7 deals with basic epistemological issues involved in the attempt to implement a computer model useful for the investigation of social networks. The issues as presented seem to illustrate clearly and to represent a realization of the need to focus on modeling *activity* as the necessary approach to a better understanding of complex phenomena. The authors use their experience in attempting to describe and simulate psychosocial behavior to argue that in dealing with such phenomena "aims of theory unification should be replaced by aims of coordination of competitive and complementary theories." The modeling approach reported by the authors represents a significant development toward this goal.

Chapter 8 provides an interesting systems-based analysis of one of the fundamental tools available to the social scientist, the sociological interview. The reported developments, which essentially represent a general systems investigation whose object of investigation is itself a tool, again underscore the recognition of the importance of adequate methodological procedures.

Chapter 9 presents a promising approach for dealing with complex, multiobjective decision situations. The approach and the model described in this chapter, by implementing a view and perspective on problem situations, stress the need to deal with consistent wholes and integrate heuristic coordination rules to define an operational global adaptation process. The result represents a valuable alternative approach to multicriteria decision making and further illustrates the need to integrate abstract results in context of the general study of problem-solving processes.

The direct significance and potential of a systems orientation for the study of complex social phenomena is well illustrated in this last section. We hope that these papers motivate further research and development along these lines.

7 SOCIAL NETWORKS AND MULTILEVEL STRUCTURE: *System Description and Simulations*

Stein Bråten, Eivind Jahren, and Arild Jansen

The study of psychosocial behavior, such as communication and cognition in voting contexts, offers specific problems of system description and theory exploration. It should cover aspects of the interaction between actors, of their social network context, and of the intraactor structure that may take part in the production of psychosocial behavior. It should also permit exploration of propositions drawn from various theoretical "islands." This makes for a complexity that demands computer simulation as an aid in generating the implications of such bridges and adapted designs that allow for comparison between model systems and referent systems. If one assumes—as we do—that such referent systems obtainable from a given data source may reveal competing and complementary patterns, then a multiple approach in terms of a set of models that differ in viewpoints and levels is called for.

The following is reported from computer simulation with a set of models of a "network panel" surveyed in a Norwegian community during a referendum con-

The research reported in this paper was done at the Institute of Sociology, University of Oslo, supported by grant no. B.30.23.077 from the Norwegian Research Council for Science and the Humanities (NAVF), and grant no. B.98.00.55 from the Norwegian Broadcasting Corporation (NRK).

troversy in 1972, through the application and extension of a special systems paradigm (Bråten, 1966, 1971) that previously had been used in a number of studies of human communication and social cognition. Allowing for systems selection at aggregate (panel) level, network and dyad level, and (intra-)person level, the network panel survey data serve as a basis for implementation of three models that vary with respect to level and theoretical perspective. They are contained in a multiple simulation structure and serve in the exploration of some problems associated with shifts of definition of system, level, and time, with reference to a common data base.

MODEL SYSTEMS DESCRIPTIONS

Referent Systems

The referent systems of the models in the multiple-model structure described in this chapter are offered by a panel of respondents sampled from the Oslo area in Norway and studied in two waves prior to and one wave after the referendum on Yes or No to Norway's entering the European Common Market (EEC). Respondents are partly linked together in household and acquaintance nets (77 in all). In the first wave 241 respondents are questioned, while 209 of these are available in the third wave.

Through combining a network and a panel orientation, this "network panel" (Jahren, 1973) combines intentions involved in earlier empirical approaches to personal influence (Katz and Lazarsfeld, 1955) and social diffusion (Coleman, et al., 1957; Rogers and Svenning, 1969). It is designed according to the cognitive consistency and communication network perspective of a previous model, SIMCOM (Bråten, 1968), which is related to the Abelson and Bernstein (1963) referendum model. This has been used in a postdictory simulation (Bråten, 1976) of the EEC referendum at the national level with reference to data from public opinion and election polls.

Model Systems Levels and Viewpoints

A class of viewpoints that allows for compatible selections of systems, differentiated only with respect to degree of detailed level description, is here conceived of as sharing the same *system perspective.* The surveyed network panel may be approached in terms of systems of different levels of *structural resolution* (Klir, 1969) or *actor resolution* (Bråten, 1971) or *strata* (Mesarovic, et al., 1970), if the latter is restricted to the detail level sense as distinct from the functional strata sense, and for which it holds that movements toward finer levels involve more detailed explanation (Klir and Valach, 1967).

The levels that are to be involved with reference to the above panel are as follows:

1. The *aggregate (panel) level,* which is the grossest level for description of distribution of states and behavior of the population of persons and which disregards their mutual couplings—if any—but may involve the conception of the panel as an aggregate actor;
2. The *network and dyad level,* which allows for specification of the panel population structure in terms of strongly connected actor pairs (household dyads) and weakly connected persons (acquaintance relations), as well as unconnected ones, and permits descriptions in terms of degree of balance of dyads and nets;
3. The *(intra-)person level,* which allows for specification of the intrapersonal structure and mechanisms and action programs that make up interaction and that require actor-coactor participation to be completed.

Thus, the network panel may be approached through a number of simulation models that share the same perspective [e.g., that of cognitive consistency, (Abelson et al., 1968)], but vary with respect to the level(s) at which this principle is applied. One model may involve mechanisms operating at the intrapersonal level, while another may restrict its operations to the interpersonal or network level. While such mechanisms are allocatable to different levels, they may be compared in terms of system behavior at the same level (e.g., the dyad and network level). However, the network panel data base also allows for the construction of models differing in systems perspective. A model built from a socioeconomic-entity perspective represents a viewpoint incompatible with models representing consistency and communication-network viewpoints. In the former perspective, panel respondents become entities in an aggregate, capable of state changes as a function of socioeconomic background variable values, but they lose their characteristic of being information-processing members of communication nets. Still, state changes generated by such a model may be compared at the aggregate (panel) level with those of competing models and of the empirical referent system.

Such a socioeconomic-entity perspective is represented by one of the models—model III—reported below, while the other two share a cognitive-consistency perspective. According to the theoretical principle of the latter, states of attitudinal inconsistency in and between persons and social groups and cross-pressure of opinions create a strain in the system. This strain affects communicating and information-processing behavior and tendencies of attitudinal switch and withdrawal during attempts at "resolving" the experienced inconsistency (Abelson, et al., 1968; Bråten, 1977; Cartwright and Harary, 1965). The simpler of the two models sharing this perspective is model II, which is constructed from a network and dyadic balance viewpoint and is influenced by a graph-theoretical version of cognitive consistency theory (Cartwright and Harary, 1965). Model I encompasses intra- and interpersonal mechanisms for consistency and communication processing in line with some previously developed models (Bråten, 1968, 1976; Abelson and Bernstein, 1963).

The three models, I, II and III, as well as the multiple simulation structure containing them, are written and implemented in the SIMULA language (Dahl and Nygaard, 1965; Birthwistle et al., 1973). The global structure, labeled MULTISIM, is being constructed to offer opportunities to explore some inherent ambiguous and pluralistic aspects of scientific explanation. This is recognized by Næss (1972) in his approach to philosophy of science, in the multilevel and multimask orientation in the general systems methodology of Klir (1969, 1975), in Merton's treatment of social structure theories (1975), and in Bråten's systems and simulation approach to psychosocial behavior (Bråten, 1971, 1973, 1977; Guetzkow et al., 1972).

Application of the Modeling Paradigm

The paradigm used in the present study has been developed by Bråten (1971). Parts of it have previously been applied in the following capacities: in an approach to market communication (Bråten, 1966); in the network panel design providing the data source for the present context (Jahren, 1973); in theory-explorative simulations of mass communication and cognitive consistency (Bråten, 1968); in a referendum simulation at the national level (Bråten, 1976); in a study of moral dilemma processing dyads (Bråten, 1972); in simulation of a student-teacher interaction (Norlén, 1972); and in general business and economic modeling (Bråten and Norlén, 1973).

These models have been constructed in terms of intra- and interactor action programs. Their building blocks are processes of shift and activation of choice, adaptation and action programs, processes of orientation and image activation, and processes of manifest signal transfer. As illustrated in Figure 7.1, these processes may be conceived as occurring in three different fields—the program field, the orientation field, and the sign(al) field, affecting and being affected by sets of (field) state variables in and between actors, mediated by internal and externalized messages. As the execution of any action program—such as message production or reception—on the part of an actor presupposes the activation of a co-actor program in order to be completed, these models thereby involve at

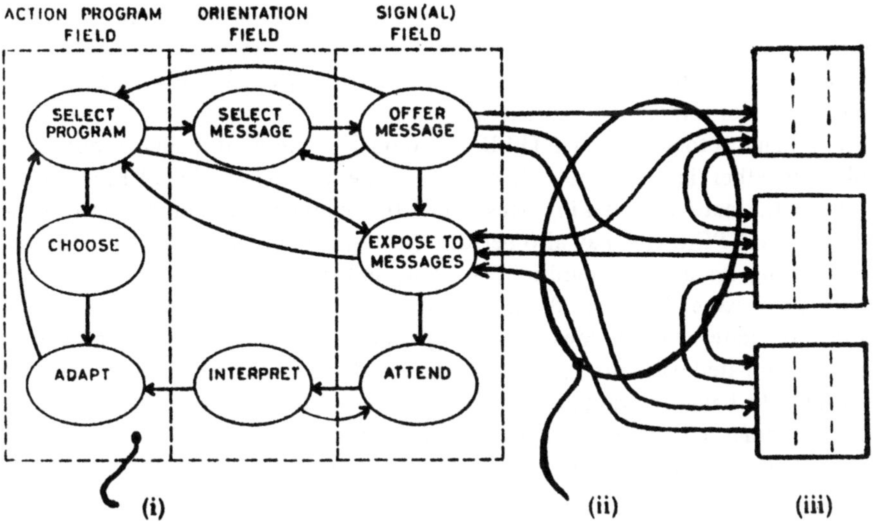

Figure 7.1. Type of Gross Actor and Interactor Structure according to Modeling Paradigm: (i) Intraactor Organization, (ii) Interactor Message Flows, (iii) Personal and Organizational Co-actors in Network

least a context of a human dyad or networks transcending the actor conceived as a monad. The cell for concurrent execution of an actor and co-actor program requires a programming language that allows for data- and program-carrying (actor) objects that may exist in parallel and allows for description at several levels of (actor) resolution. These system-description and computer-operational desiderata have made SIMULA almost ideally suited to the computer application of the paradigm. In addition, the principal designers of SIMULA, Dahl and Nygaard, have become highly influential in this application (Dahl and Nygaard, 1965; Birthwistle, et al., 1973).

STRUCTURE OF MODEL I (NETPERSON MODEL)

The structure, programs, and mechanisms of the most complex model with reference to the network panel are closely related to those of the SIMCOM model. (This model should not be confused with another one bearing the same label and having been independently developed by members of the Department of Geography, Northwestern University. Both "Simcom" models are briefly discussed in Guetzkow et al., 1972.) The SIMCOM model is constructed from a perspective that combines propositions from theories of cognitive consistency and (personal and mass) communication. The model, involving at least one co-actor, operates in terms of actors that are coupled systems, capable of executing programs that produce state changes in their internal or external environment as part of interaction sequences.

The main actors in the NETPERSON model are network panel respondents linked together in social nets. Instances of the actor-class NETPERSON may perform the actions of taking pause, initiating contact, and complying to contact. The NETPERSON may also perform (intra-) personal state changes in terms of adjusting evaluation, cognition, interest, and consistency in relation to the controversy issue. Other kinds of actors and message sources are incapable of intraactor state changes. Messages being composed, exchanged, and processed in the simulated systems carry evaluative and informational content and are restricted to being relevant to the yes or no issue. *Actions* involve state changes in the interactor system through the execution of some program, selected from the set that defines the action alternatives of the actor class.

Gross Coupling Characteristic

Action programs are composed of processes. They are defined as directional events occurring within and across different actor-system fields and containing mechanisms that operate on input, throughput, and state variables, producing values on throughput, state, and output variables. The processes used in the NETPERSON model to define the various intra- and interpersonal actions are as follows:

1. *Select action program* (i.e., decide which action is to be performed next and at what time, including the action of doing nothing through holding pause);
2. *Select message* (i.e., compose evaluative and informational content);
3. *Select and send to co-actor* among available net members or to dummy co-actors outside net, without any guarantee of compliance on part of the activated co-actor;
4. *Expose* and *attend* to message from some active co-actor or message source;
5. *Interpret* co-actor's message offer, including noisy distortion of its content;
6. *Adapt* internal actor-state variables, such as current state of interest, commitment, knowledge, evaluation, and cognitive (in)consistency.

The model mechanisms (operating characteristics) are called on during these processes and contain the various mechanisms, including operating characteristics, according to cognitive consistency theory. Thus, for instance, the actualized state of affective-cognitive inconsistency results in systematic change in overt and covert communication and information-processing behavior of the actor experiencing the inconsistency. He becomes more active in his communicative behavior through more frequent action program activation on his part, more discriminative in his co-actor selection, more selective in his message attention, and more varied in his distortive interpretative behavior, as compared with his covert and overt behavior when being consistent. If none of the above "inconsistency resolution modes" releases him from his state of inconsistency, he will resort to a switch of his evaluation preference or withdraw to an uncommitted state (responding "do not know" or not responding or becoming a "home-sitter" at the date of voting). Such inconsistency is experienced by the person if he evaluates positively both the yes-issue and the no-issue or has knowledge stored in favor of both sides; it may be caused by being subjected to cross-pressure (i.e., to sources promoting both sides) or through being engaged in communication with a co-actor or source promoting the opposing side, or merely through being a member of a network or an aggregate of persons dominated by the color opposite that of the actor in question.

Thus, the model mechanisms utilize actual values of field-state variables in the actor and as reflected at grosser levels and produce state changes at the (intra-) person level. This action in turn affects the actual state at grosser levels, indirectly through communication, and directly through each person's being a part of a more encompassing system. Thus, intrapersonal consistency, interpersonal and network balance, and aggregate (panel) proportions of yes-, no-, and uncommitted persons affect and are affected by the operation of the NETPERSON model mechanisms. The couplings between the processes containing these mechanisms are depicted in Figure 7.2.

Model Structure in Terms of Declarations in SIMULA

The SIMULA language is a proper extension of ALGOL 60. The concepts of blocks (BEGIN, END enclosures) and procedures are extended to CLASS objects. An activity may be formally declared as

```
PROCESS CLASS ACTIVITY (⟨parameter list⟩);
   ⟨formal parameter declaration⟩;
   BEGIN
   ⟨declaration of attributes (variables and procedures)⟩;
   ⟨program statements⟩;
   END;
```

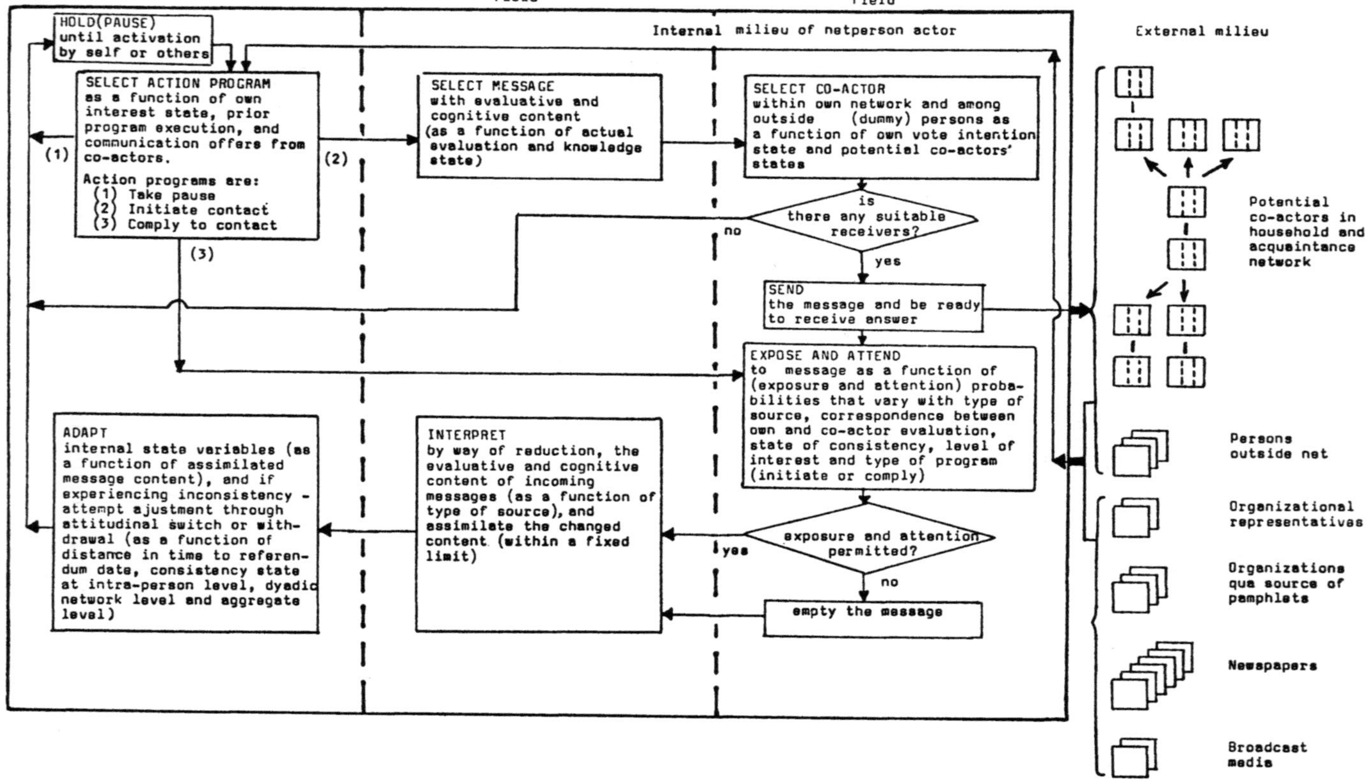

Figure 7.2. Gross Representation of Coupling Characteristic of the NETPERSON Model I with Reference to the Empirical Network Panel (Bråten, 1968, 1976)

Classes can be prefixed, defining subclasses, such as

ACTIVITY CLASS ACTOR,

whereby all characteristics of ACTIVITY are inherited by ACTOR. During program execution many "instances" of the class object may be physically generated and exist at the same time. To refer to and access each (distinct) of them, a reference concept REF is available, for example,

REF (ACTOR) COACTOR,

which allows for generating and accessing the "instance" referred to:

COACTOR:- NEW ACTOR (⟨actual parameter list⟩);
IF COACTOR. YESPERSON THEN CONTACT; . . . . ;

since YESPERSON is a Boolean attribute of ACTOR, made available through the genitive-type "dot" notation (.).

The concept VIRTUAL is useful in defining class-subclass hierarchies. Procedures and label attributes may be formally defined but left unspecified at an outer level. This allows for different actual definition of the virtual elements in the distinct subclass declarations.

The simulation facilities, such as TIME concept, event scheduling, and activate and passivate procedures, are built into the language as SIMULA-defining process classes and are thereby made available to all implemented, program-defined subclasses.

The Actor Class. The core of the NETPERSON model is a set of (NETPERSON) actors, which exchanges and receives information and evaluations through personal interaction and mass media communication. In the model, the declaration of

ACTIVITY CLASS ACTOR; . . . . . . ;

enables all ACTOR and ACTOR subclass members to behave as dynamic, time-consuming objects. They are offered SIMULA procedures for activation and passivation, and "sleeping" states until being awakened by co-actors who want to start interaction, by message sources such as newspaper editions, or by their own "alarm clock," which is the result of a HOLD(PAUSE) statement.

Actors are of different types. Their common traits in terms of state variables, action program labels, and switch between programs are declared below:

```
ACTIVITY CLASS ACTOR;
   VIRTUAL: PROCEDURE START;
            LABEL TAKEPAUSE, INITIATECONTACT,
            COMPLYTOCONTACT;
   BEGIN
      BOOLEAN YESPERSON, NOPERSON, NEUTRALPERSON;
      INTEGER PROGRAM, PAUSE;
      INTEGER PROCEDURE INSPECTSTATE; ;
      PROCEDURE CHANGESTATE; ;
      REF (ACTOR) COACTOR;
      REF (MESSAGE) INMESSAGE, OUTMESSAGE;
      SWITCH ACTIONPLAN(PROGRAM) := TAKEPAUSE,
                                    INITIATECONTACT,
                                    COMPLYTOCONTACT;
      .
      .
      .
      .
   END;
```

There are three ACTOR subclasses—NETPERSON, OUTSIDENETPERSON, and ORGREPRESENTATIVE—each with its distinct action programs pattern that is dependent on the actor's position in the social structure.

NETPERSONS correspond to respondents of the empirically surveyed network panel and are located within the model-implemented networks. They are the most completely equipped actors in terms of operating characteristics (mechanisms).

OUTSIDENETPERSONS are located outside the networks, but within the NETPERSON's available social environment. Empirically they can be identified as family members and friends, but we do not allow for access to their internal organization (i.e., the variables describing their *Innenwelt*). Consequently, they are not model persons able to handle, store, and adapt messages. Their action program capabilities consist merely in registering contact trials from others (NETPERSONS) and sending empirically predefined messages.

ORGREPRESENTATIVES denote person actors acting on behalf of political parties and ad hoc organizations. Similar to OUTSIDENETPERSONS, they are modeled without internal message-processing capabilities. Consistent with the notion of organizational representatives as "agents" promoting one unrevisable position, ORGREPRESENTATIVES may search for candidates to influence or act on within an empirically defined set of NETPERSONS.

Message Generation. Message classes are organized partly in parallel to the activity class structure. They exist as data carriers available to the actor and mass media activity classes. Different types of messages reflect actors and activities implemented in the model.

Mass media messages are "newspapermessage" (comprising six newspapers), "pamphletmessage" (which takes care of the mass media activities of the ad hoc organizations), and "broadcastmessage" (covering radio and television).

Personal messages are "orgrepmessage" (from representatives of political parties and ad hoc organizations), "outnetmessage" (from persons belonging to the NETPERSON's social environment but not as part of his social network qua panel respondents), and "netmessage" (from NETPERSONS).

Only netmessages are being changed during the simulation run, as a function of the producing NETPERSON's actual cognitive and evaluative states. All messages are characterized by values on evaluative "yes-content" and "no-content" and "cognition-content." The former two variables reflect the degree to which the message is tilted toward one or the other side of the two-issue alternative, while the latter reflects the ability of the source to communicate relevant information.

The Actor Class NETPERSON. NETPERSONS have a fully developed internal organization. They are able to carry through a complete interaction sequence with other NETPERSONS as well as to let themselves be influenced by mass media, persons outside the nets, and organizational representatives. Their initial states are given according to the reported states at t_0 of the corresponding respondents. The mechanism-empty structure of the subclass NETPERSON declaration is as follows:

```
ACTOR CLASS NETPERSON;
   BEGIN
   REAL YESVALUE, NOVALUE, KNOWLEDGE, INTEREST;
   BOOLEAN SINGLE, ORGMEMBER;
   REF (NETPERSON) HOUSEMEMBER, AQUAINTANCE;
   PROCEDURE SELECTMESSAGE; ;
   PROCEDURE SELECTCOACTOR; ;
   PROCEDURE SEND; ;
```

```
        PROCEDURE EXPOSEANDATTEND; ;
        PROCEDURE INTERPRET; ;
        PROCEDURE ADAPT; ;
        .
        .
        .
(1)     TAKEPAUSE:          PROGRAM := 2; HOLD(PAUSE);
                            GOTO ACTIONPLAN(PROGRAM);
(2)     INITIATE CONTACT:   SELECTMESSAGE; SELECTCOACTOR;
                            IF COACTOR == NONE THEN GOTO
                            TAKEPAUSE
                              ELSE BEGIN
                                        SEND; EXPOSEANDATTEND;
                                        INTERPRET; ADAPT;
                                      END;
                            GOTO TAKEPAUSE;
(3)     COMPLYTOCONTACT:    EXPOSEANDATTEND; INTERPRET;
                            ADAPT;
                            IF COACTOR == NONE THEN GOTO
                            TAKEPAUSE
                              ELSE BEGIN
                                        SELECTMESSAGE; SEND;
                                      END;
                            GOTO TAKEPAUSE;
        END NETPERSON;
```

According to the above gross declaration, a NETPERSON actor may carry out one of three alternative action programs: (1) TAKEPAUSE, (2) INITIATECONTACT, or (3) COMPLYTOCONTACT. He is always performing one of them, depending on (1) his own internal states (interest level and consistency state) and (2) contact trials from the environment. Each action program characterizes types of processes or action sequences:

TAKEPAUSE is quite simple. It implies that the NETPERSON abandon any further activity until he is activating himself or is being activated by others (person actors or mass media).

INITIATECONTACT involves activation by the NETPERSON himself to initiate interaction with some other person. In this action program the NETPERSON selects a message, whose evaluative and cognitive content is a reflection of his evaluation and knowledge state. From a list of possible co-actors in his social environment, where some are more preferable than others, he selects one and sends the message to him before passivating himself.

COMPLYTOCONTACT activates the NETPERSON as a result of initiating efforts by co-actors or message sources and involves mechanisms regulating exposure and attention, interpretation and adaptation.

Execution. When the NETPERSON model starts execution, all actor and mass media objects are generated. To each empirical respondent corresponds a class "instance," with identical initial attributes and state variable values. In a similar way, mass media objects (newspapers, organizations, and so on) are created with corresponding edition frequency. They are all activated at once. All NETPERSONS will go to "pause" for a while, and the mass media start creating and distributing new message instances by the statements

```
NEXT MESSAGE :- NEW MESSAGE;
SUBSCRIBER.INMESSAGE :- NEXT MESSAGE;
ACTIVATE SUBSCRIBER;
GOTO TAKEPAUSE;
```

Interpersonal communication will be initiated through netpersons entering an initiate-contact program (1). Although the procedures that make up the contact action programs are called on and executed sequentially in real-world machine time, model-system time is regulated by the dyadic interactive perspective and held constant during a computer-executed sequence that simulates parallel exhange of messages. Thus the respective actions of two interactors may be regarded as occurring concurrently. The physical object blocks that represent the various actors switch between being in active and in passive phases.

When a NETPERSON who is the current actor has activated himself and entered upon an INITIATECONTACT program, he selects a message by calling on the SELECTMESSAGE procedure, after which SELECTCOACTOR becomes activated. The latter procedure involves his search for a suitable co-actor among the members of his network, or outside the panel if he is a monad. Let us assume that the actor has found a co-actor to forward his message to, through the SEND procedure. Before the occurrence of any exposure to possible return messages from the co-actor, the latter takes over as the currently active. He has been activated by the actor and enters the COMPLYTOCONTACT program. EXPOSEANDATTEND is the first procedure then called upon. If the co-actor is willing to engage in conversation, he will return a message. In any case, he will in turn activate the initiator and passivate himself.

We have just illustrated an important feature of the model structure, made possible by the quasi-parallel processing abilities of the programming language used—that is, two instances of an actor class may coexist and interact, and between two active phases belonging to one of the actors, active phases belonging to the co-actor may occur.

MODEL II (NETBALANCE MODEL)

Model II is constructed from a dyad and network balance viewpoint, inspired by a graph theoretical version of Heider's theory (Dahl and Nygaard, 1965). The operating characteristics or model mechanisms are activated at network level. They utilize the actual state of each net in terms of balanced and unbalanced dyadic relations and produce state changes in nodes of the net. Each node represents a person respondent with the possible state of 1 (yes-person), 2 (no-person), and 3 (neutral—qua withdrawn or unavailable for response). Unlike model I, intrapersonal variables are not used, nor are action or interaction programs evoked on part of the persons. They are mere network nodes without "internal life." Mutual influence is only indirectly reflected in the operation of the net balance mechanism, which utilizes a simple concept of dyadic balance. Each directly related pair of nodes is compared in terms of its actual states. For the relation to be classified as balanced, both of them have to be committed persons of the same vote (intention)—that is, both being yes-persons or both being no-persons. All remaining dyadic relations are classified as unbalanced. The integer variables BALREL and UNBALREL give the respective numbers of balanced and unbalanced relations in a given net. Being considered more strongly connected than acquaintance pairs, any household pair of nodes is made into a cycle through having their signed relation doubled. The actual value of the network state variable NETBALANCE of type real is

NETBALANCE := BALREL/(BALREL + UNBALREL).

If the actual value equals 1.0, the probability of state change in one node is .05.

If change is to be effected as outcome of this NETSHIFT drawing, the actual state of one of the net members drawn at random is changed into uncommitted (neutral). Otherwise, if there is at least one yes-person in a no-dominated net or vice versa, a "withdrawal" procedure into uncommitted (neutral) is effected for one such person, drawn at random if there are several "outsiders."

If the actual value of NETBALANCE equals .00, which is the maximum degree of unbalance for any net, then the actual state of uncommitted members, if any, is changed into a state of commitment—that is, into yes- or no-person, dependent on the color of the net dominance. If the net is not a yes- or no-dominated net, then the procedure AGGREGATEINFLUENCE is called upon, whereby the direction is determined by the outcome of random drawings that are a function of the yes- and no-proportion of the aggregate (panel) at current time. This procedure is also utilized by model instances of monadic respondents of the panel. Figure 7.3 presents the model mechanisms in the form of a gross flowchart description.

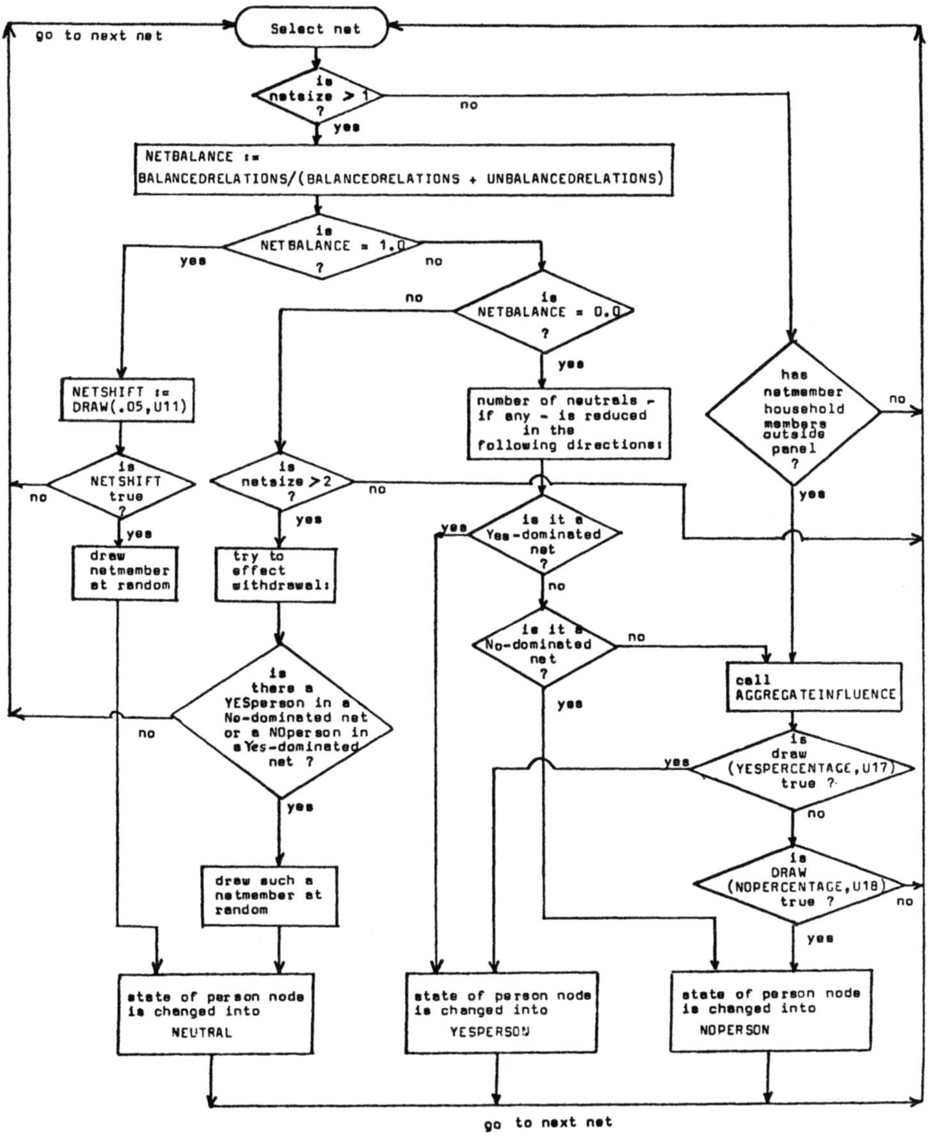

Figure 7.3. Flowchart of the NETBALANCE Model II

MODEL III (SOCIOEC-INDEX MODEL)

This model is the least complex in the model set. It represents a system perspective whereby most system properties are erased. It is constructed from a socioeconomic entity point of view, but in too crude a manner to do justice to such a modeling viewpoint.

Three postulated sets of state transition matrix values for the person entities as three-state systems are used, conditional to the position of each person on a socioeconomic index. The index is based on age, sex, education and income, and allows for low, medium and high position, with each position associated with a set of state transition values. The procedure POSITSTATE(NEWSTATE) sets the next state of the person entity as one of three (1, 2, 3—for yes-, no-, or uncommitted person, respectively) through the use of the real auxiliary variable R and the declared

INTEGER ARRAY TRANSMATRIX(O:MAXECINDEX, 1:3, 1:3).

The model proceeds as follows:

```
COMMENT
find socioeconomic index position of this person,
select associated matrix,
define present state of this person;
R := UNIFORM(0, 1, U6);
WHILE R > TRANSMATRIX(SOCIOECINDEX,STATE,NEWSTATE) DO
   POSITSTATE(NEWSTATE);
COMMENT
which according to the matching interval between the resulting value on R and
   the matrix sets the new state for the person entity.
```

The above sequence is then repeated for each panel member. This model is blind to the person's being a member of any network or more encompassing aggregate qua sources of influence, but takes the person's initial state into account.

MULTIPLE SIMULATION MODEL STRUCTURE AND BEHAVIOR

From a series of empirically linked simulation model studies has emerged a basis for extending the paradigm applied to these models into a multiple-oriented approach to psychosocial behavior (Bråten, 1971, 1972, 1977). This could be consistent with the desideratum concerning shifting of masks and comparison of models in general approaches to systems and simulation, such as represented by Klir (1969) and Zeigler (1976). In the sketch below of the MULTISIM structure, conceptual terms adapted toward the terms used in a more general systems sense have not been attempted.

The same psychosocial phenomena, such as voting and communication in a network panel, as studied above, may be approached in terms of a plurality of models of various levels and viewpoints. Rather than utilizing a special systems approach that unifies or synthesizes diverse models so that they lose their distinctive characteristics, one may strive toward a fairly "empty" structure that allows for the concurrent housing of, and shifting between, members of such a model set. Such a MULTISIM structure, currently being implemented by SIMULA, allows for the activation of the competitive models described above, as well as of alternatives to these, with reference to the network panel survey data base. MULTISIM is being constructed to offer opportunities of exploring and exploiting the inherent ambiguous and pluralistic aspects of scientific explanation. Its structure is illustrated in Figure 7.4.

The MULTISIM structure has been crudely implemented in SIMULA, in that so far many of its procedures are dummy or empty procedures. MULTISIM serves the functions of comparison, regulation, and shift between systems of different levels and perspectives.

Meta-Objects

MULTISIM thus needs to be a structure capable of operating at several levels and from several perspectives. Its components are ideally able to observe, describe, and record its own behavior (the observing model creating an image of itself). Meta-objects are introduced as part of the structure. They study the various images and determine if changes in action patterns or shifts of system perspective are called for. Furthermore, they regulate the activity of the various level objects, qua subclass of META CLASS MULTILEVEL.

```
PROCESS CLASS META;
   VIRTUAL: PROCEDURE STUDYPROTOCOL,STUDYEMPIRI,
                              TESTSHIFT, IMPLEMENTSHIFT;
            LABEL: REST, EXAMINE, REGULATE;
   BEGIN
   .
   .
   .
   END META;
```

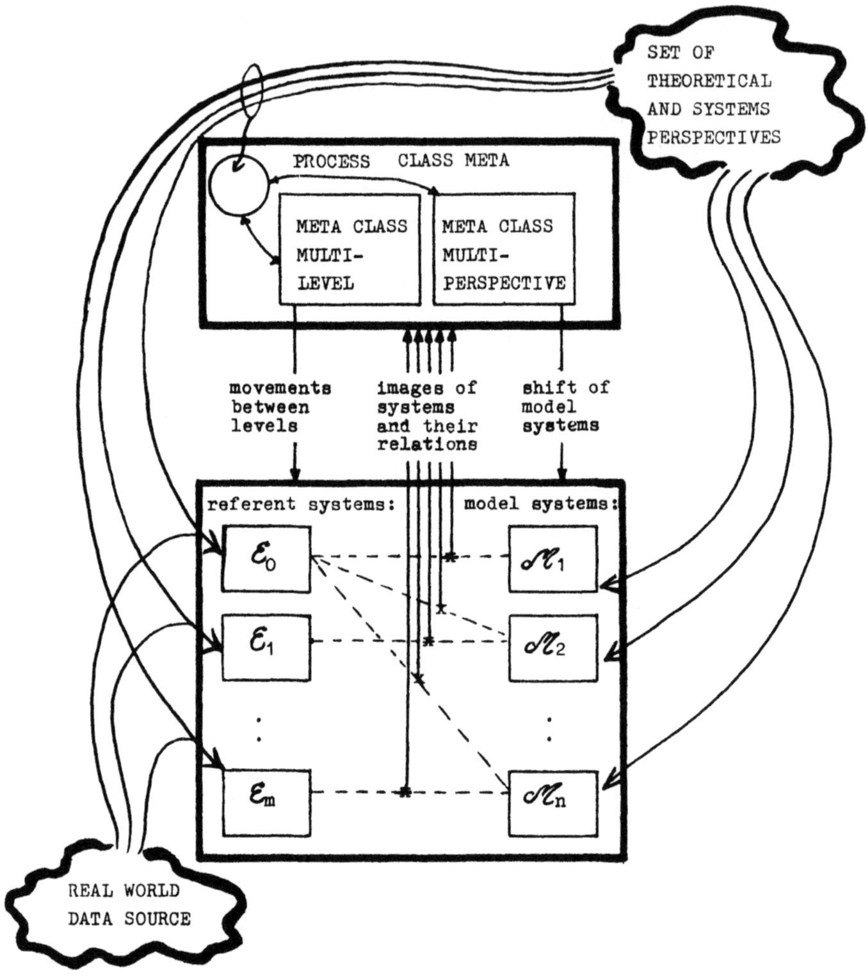

Figure 7.4. The MULTISIM Structure in Line with a Special Systems and Simulation Paradigm (Bråten, 1971) and Partly with General Systems (Simulation) Approach (Klir, 1969; Zeigler, 1976)

```
META CLASS MULTILEVEL;
   VIRTUAL: PROCEDURE REPORTSHIFT;
   BEGIN
   PROCEDURE TESTSHIFT; ;
   PROCEDURE IMPLEMENTSHIFT;
      BEGIN
```

```
    SWITCH LEVEL := AGGREGATE,NETWORK,DYAD,INDIVIDUAL;
    END IMPLEMENTSHIFT;
  .
  .
  .
  END MULTILEVEL;
```

The levels are implemented in terms of process subclasses, and are

MULTILEVEL CLASS AGGREGATELEVEL; ;

which records actual aggregate (panel) state, tests for changes in aggregate proportions and reports to protocol;

MULTILEVEL CLASS NETWORKLEVEL; ;

which records state variables at the network level, counts number of yes-, no-, and uncommitted persons in the various nets, computes degree of balance, and reports to protocol;

MULTILEVEL CLASS DYADLEVEL; ;

which describes each household and acquaintance dyad as balanced or not balanced, tests if a state shift is possible, and reports actual shift to the protocol;

MULTILEVEL CLASS INDIVIDUAL LEVEL; ;

which records actual person and intraperson states in terms of person state field variables and changes accordingly the values of the gross Boolean state variables YESPERSON, NOPERSON, and NEUTRALPERSON, and reports this to the protocol.

When the different levels are represented as PROCESS CLASS (subclass) objects, all levels may be activated at the same time and execute their action programs in parallel, or they may compete and exclude each other, depending on the regulating procedures. This reflects the multilevel structure: the different level objects behave partly independent of each other and operate at distinct variables, but they are also mutually influential as changes in person states will affect dyad, network, and aggregate states, which in turn provide a changed context for objects at finer levels.

Perspectives. To allow for competing perspectives, *P* or *Q*, for example, in the case of the three models I, II, or III, where I and II represent a cognitive consistency perspective (*P*), while III a socioeconomic perspective (*Q*), a subclass of META regulates shifts between *P*- and *Q*-perspectives.

```
META CLASS MULTIPERSPECTIVE;
  BEGIN
  INTEGER PERSPECTIVE;
  BOOLEAN SHIFT;
  SWITCH MODEL := PPERSPECTIVE, QPERSPECTIVE;
  REST:       HOLD(PAUSE);
  EXAMINE:    STUDYPROTOCOL; STUDYEMPIRI; TESTSHIFT;
              IF SHIFT THEN GOTO REGULATE ELSE GOTO REST;
  REGULATE: IMPLEMENTSHIFT; GOTO REST;
  PROCEDURE TESTSHIFT;
    BEGIN
    COMPAREAGGREGATE;
    IF AGGDEVIATION > AGGTHRESHOLD THEN SHIFT := TRUE;
    COMPAREINDIVIDUAL;
    IF INDIVIDUALDEVIATION > INDTHRESHOLD THEN SHIFT :=
      TRUE;
    END TESTSHIFT;
  PROCEDURE IMPLEMENTSHIFT;
    BEGIN
    GOTO MODEL(PERSPECTIVE);
    PPERSPECTIVE: TERMINATEQ; INITIATEP; ACTIVATE
                  MULTILEVEL; ACTIVATE IMAGE; GOTO REST;
    QPERSPECTIVE: TERMINATEP; INITIATEQ; ACTIVATE
                  MULTILEVEL; ACTIVATE IMAGE; GOTO REST;
    END IMPLEMENTSHIFT;
  END MULTIPERSPECTIVE:
```

Model Set

Model I. As partly indicated by its label NETPERSON, model I is constructed from an interpersonal communication and cognitive consistency viewpoint. Its mechanisms are activated at the (intra-)person level and utilize person-, dyad and network-, and aggregate-state variable values. In turn, they directly and indirectly produce changes in these values.

Model II. This NETBALANCE model is constructed from a network balance viewpoint and is influenced by a graph-theoretical version of cognitive consistency theory. The mechanisms are activated at the network level and utilize the actual state of each net in terms of balanced and unbalanced dyadic relations. They produce state changes in nodes of the net.

Model III. In this model three postulated sets of state transition matrix values for the person entities as three-state systems are used and are conditional to the position of each person on a socioeconomic index (hence the label SOCIOEC-INDEX model). The index is based on age, sex, education, and income, and allows for low, medium, and high position.

Referent System and Model Systems Behavior

Gross Couplings and State Transition Structure. MULTISIM records system behavior in the following forms:

1. *State transitions* as part of ST-structure (Klir, 1969) between t_0 and t_1, and between t_1 and t_2 (see Table 7.1 on referent system behavior);
2. *Graphical description* of couplings of persons qua elements (part of a gross UC-structure [Klir, 1969]), with actual states of person nodes at different time points in terms of yes-state (+), no-state (-), and uncommitted (0);
3. *Aggregate proportion* of states generated at different time points (see Figure 7.5).

MULTISIM also records other kinds of *snapshots*, for example, of intraperson states and contents of messages being paid attention to (in order to provide a working image of Model I).

Qua members of MULTISIM the three models have only been allowed a first run with reference to a reduced panel of 131 members, due to a problem caused by core memory limitations of the computer used (DEC SYSTEM 10). However, model I has also been executed in runs with reference to the whole network panel, and over such periods of time that fairly stable aggregate states are generated.

The state transition structure of the reduced referent panel at the aggregate (panel) level is represented in Table 7.1. It gives transitions (1) from the first to the second wave (before the referendum date) and (2) from second to third wave (t_2), which is after the referendum date. Yes-person state is denoted by +, no-person state by -, and uncommitted or unavailable by 0. Table 7.2 shows the

Table 7.1. State Transition Structure of Reduced Empirical Network Panel of 131 Respondents in Periods (a) t_0-t_1, and (b) t_1-t_2, in Terms of the States of Being Yes-Person (+), No-Person (-), and Uncommitted or Unavailable for Response (0)

		(a) t_1 +	–	0			(b) t_2 +	–	0
t_0	+	.92	.00	.08	t_1	+	.96	.01	.03
	–	.13	.61	.26		–	.04	.89	.07
	0	.13	.25	.63		0	.12	.19	.69

initial proportions of yes-, no-, and uncommitted persons, and as generated at t_1 and t_2 by referent (*a*) and model systems I, II, and III (*b–d*).

Couplings and person states of three networks (numbers 5, 15, and 22) in the panel are shown in Figure 7.5, where they are compared to states of the corresponding networks generated by simulation models I, II, and III. Each model represents different viewpoints and levels. In the figure yes-person is denoted by +, no-person by –, and uncommitted person by 0. Coupling and initial states are identical for referent system and the three model systems.

Model Status. So far, model I may at least be considered not to show pathological behavior, which is not surprising considering it has not been built from scratch but draws heavily on a related model that has been used fairly successfully with reference to the same referendum at the national level. However, none of the three models in their initial versions succeeds in serving as an adequate

Table 7.2. Yes-Persons (+), No-Persons (-), and Uncommitted Persons (0) at Time t_0, t_1, and t_2 in Referent System and Model Systems

	Time t_0			*Time t_1*			*Time t_2*		
State	+	–	0	+	–	0	+	–	0
a. empirical	.59	.29	.12	.60	.21	.19	.60	.23	.17
b. model I	.59	.29	.12	.63	.34	.03	.63	.31	.06
c. model II	.59	.29	.12	.69	.23	.08	.63	.24	.13
d. model III	.59	.29	.12	.58	.28	.14	.48	.34	.18

Note: Percentage of number of panel members reduced to 131.

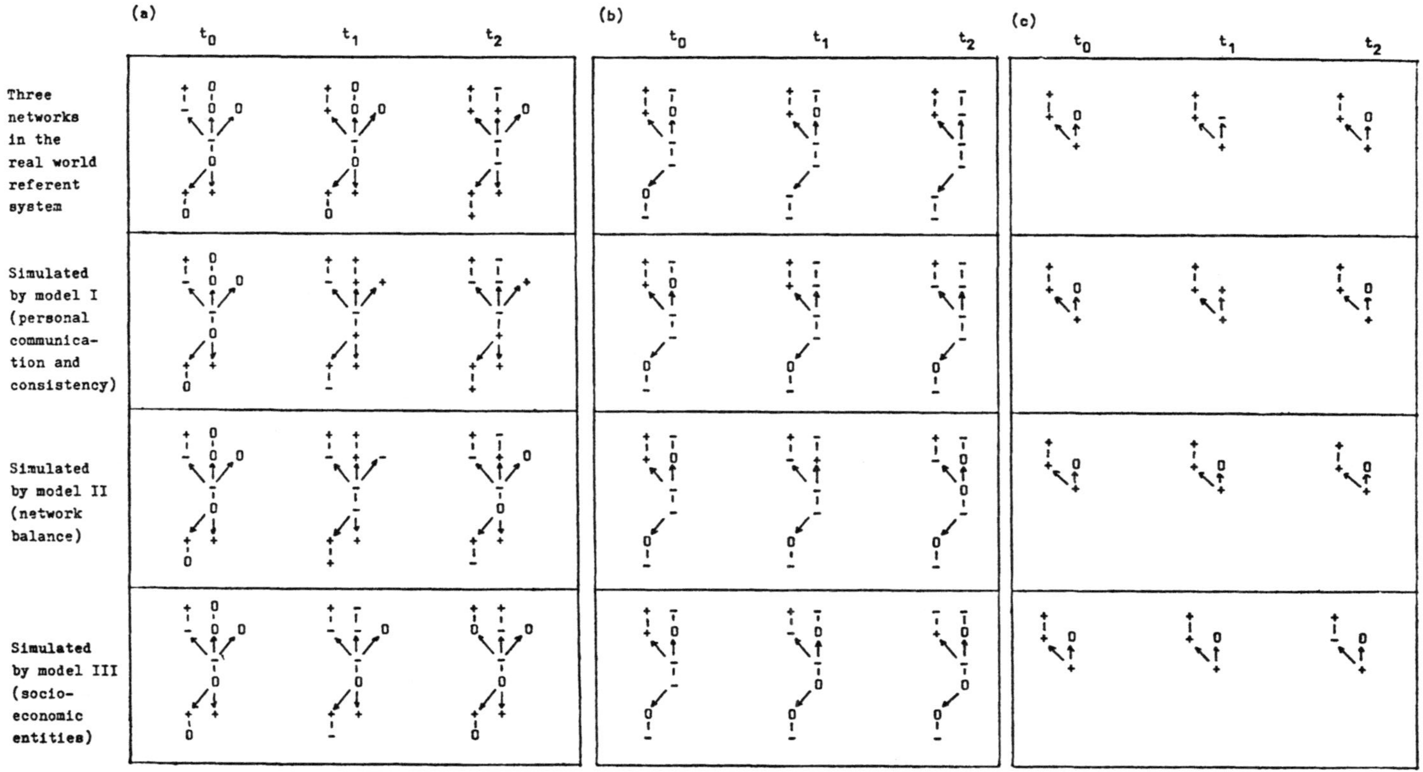

Figure 7.5. Couplings and Person States of Three Networks (Numbers 5 [a], 15 [b], and 22 [c]) in the Empirically Surveyed Panel at Times t_0, t_1, t_2, and States Generated by Simulation Models I, II, and III

descriptive or explanatory vehicle. Based on feedback from experimentation with the aid of MULTISIM, the following model modifications are anticipated:

1. The intra- and interpersonal consistency and communication model will probably be replaced by one of the same type that is capable of discriminating between different categories of "neutrals,"—for example, between those that for their own reasons withdraw from the field and those that are being made into nonrespondents by "external" causes—and that also takes care of individual time clocks in a more convincing manner.
2. The household and acquaintance net balance model may be replaced by a version that is more time sensitive with respect to proximity to the final date of decision.
3. The socioeconomic index entity model, incapable of offering anything with respect to explanation of state changes of nodes in networks in second period, may be replaced by one that is socioeconomic background *and* environment sensitive.
4. A fourth model should be added at a finer level, making use of a socioeconomic exchange perspective—nodes exchanging rewards and punishments according to status and state.

Feedback to Problems Associated with a Multiple Approach

In spite of the fact that mismatch between referent systems behavior and model systems behavior has been revealed through the first runs with the models, and modifications are required, their implementation qua local members of a global MULTISIM structure with reference to a common data base has contributed to actualizing the following general problems:

1. Problems associated with the fact that certain *masks*, that is, fixed sets of sampling elements (Klir, 1969), need be chosen in the establishment of a data base. These masks and the applied systems definitions will not only affect the variety of the set of potential models allowed to be constructed but also their respective chances of emerging as explanatory vehicles of reducible form.
2. Problems raised by adjusting definition of *levels* with shift of model perspective, and which then prevent a true comparison in terms of model system behavior.
3. Problems and challenges involved in relating model systems *time* definition(s) and referent system time.

4. The relation between unification and coordination of several theories or models. For a unifying structure to achieve simplicity and solvability, the unified elements would have to lose their respective identities and viewpoints. Their mere coordination as elements competing in a common field requires a theory and methodology for such coordination.

Level. As biases referred to under problem 1 and anticipation of level definitions (problem 2) are unavoidable, any attempt at carrying out a multiple approach will, in fact, be a *quasi*-multiple approach with a priori built-in priorities for individual members of the model set that will emerge. Models reflecting certain viewpoints and focusing on certain levels may be expected to show better results than others, merely as a function of the biases and the level of definitions implied in the data-gathering process. Thus, for example, given a certain period of trial runs and modifications toward an acceptable match between referent system behavior and model behavior at the network level, a model of type I will offer a "better" explanation than a model of type II, as the former does so in terms of structures and states at the intraperson level, the dyad and network level, and the aggregate (panel) level, while a model of type II is restricted to offering an explanation in terms of the grosser levels only. A model of type III, although being reduced in form, may compete with models of type I and II for referent system behavior match at the aggregate level. But model III is per definition prevented from offering explanation in terms of finer levels, as it allows for resolution only into isolated entities that do not have internal structures or belong to external couplings.

Time. As one moves from one type of model system to another, not only does the definition of levels change but also that of time. Three time points are empirically anchored in the dates of the three waves of the network panel survey, although the exact dates of response within each wave vary with the individual respondents. For models of type II and III, limited to produce two sets of state changes, given the input states at time t_0, a model system time may be adequately defined in approximate correspondence with referent system time. As for models of type I, although being executed as discrete event systems, a continuous time arrow linking t_0 with t_1 and t_2 is being implied, and model mechanisms are being activated *between* the time points that ought to be defined in correspondence with the empirically anchored time points.

Model system time, however, has to be punctuated between these points. Respondents report being exposed to various kinds of media messages "daily," "twice a week," "weekly," or being engaged in conversations on the referendum issue "last week," "a couple of times last month," and so forth. Contrasting this frequency of being active in terms of reported experience is the actual frequency of relevant radio and television programs and of promotional messages and articles in dailies and other printed media. This information may be used in defining the relative activation points and periods of pause of the various individual actors and media for the execution of the model system, but it does not relate to the pinpointing of the dates of the second and third survey snapshots.

The ideal empirical requirement of a model of type I is for time series provided by a larger number of snapshots than the three afforded in the network panel survey. In another postdictory simulation (Bråten, 1976) of the same referendum at the national level with the SIMCOM Model, and a simulation that offered a better match between model system and referent system behavior, a larger number of snapshots were provided by opinion and election polls at the aggregate (national) level.

Coordination versus Unification. Behind the development of the MULTISIM structure is the belief that aims of theory unification should be replaced by aims of coordination of competitive and complementary theories. Implementations of the above kind may then serve as more or less crude devices for exploring and trying out competitive and complementary theories in computer-operational form. Sufficiently rich data bases are required with a priori awareness of some of the complementary viewpoints to be applied. The encompassing and exploring of competitive models of various viewpoints and levels within some global structure, which take care of their coordination through shift between viewpoints and levels and corresponding model activation and passivation, may be regarded as a roundabout way of working toward theory unification. The shifting and coordination principles that have to be part of such a multiple structure replace principles and postulates of unification that erase the distinctive characteristics of the synthesized elements.

Postulates of coordination may be part of a coordination theory proper, as of that of Mesarovic and collaborators (1970), but not part of the theories being coordinated. In contradiction to this, principles and postulates serving a unifying function somehow have to be reflected properly by the unified theories. Furthermore, each of the models competing for activation as part of a multiple structure has to be a self-sufficient explanatory vehicle with a unique viewpoint not shared by any of the other models that are contained in the same global field structure. Anyone of them may do better or worse than others, but cannot be replaced by any of the others without reducing the variety of viewpoints being represented. Rejection of models and the replacement of a falsified one must be done *within* the set of models that share the same perspective, and *not* across incompatible viewpoints or functional strata.

CONCLUSION AND SUMMARY

The work with this multiple simulation approach is inspired by the following canons:

1. Any concrete phenomena of cognitive, symbolic, and social processes in man and in the context of his group activities contain programs of *parallel, dual, or multiple patterns* that compete for dominance or for being active and should thus be explained in terms of complementary models.
2. Any computer simulation model with some psychosocial or sociocultural real-world referent system has *interest-reflexive characteristics* and should thus ideally not be presented to the world unless as part of a set of complementary models reflecting competing interests and perspectives.

Underlying canon 1 is a systems philosophical assumption. Underlying canon 2 is an epistemological and political belief. Both of them require systems paradigms that allow for a variety of referent system definitions and complementary models with reference to a common real-world data source.

The first part of this chapter described aspects of a network panel interpersonal simulation model, implemented in SIMULA. It has been constructed with the aid of a modeling paradigm, originally intended as a means for facilitating integration of theoretical "islands" in a "meaning-tight" and computer-operational form, so as to allow for experimentation with interactor model systems in terms of symbolic and attitudinal intraactor and interactor processes. Gradually the aim of integration has been replaced by one of coordination, and the restriction to systems approached at various levels from a communication perspective is being lifted to allow for concurrent approaches in terms of several perspectives.

In addition to the above NETPERSON model, constructed from an interpersonal communication and consistency viewpoint, this chapter described two other models. One was a model constructed from a dyad and network balance perspective, and the other one was constructed from a socioeconomic entity perspective. Thus all refer to a common data base that is provided by a network panel survey, and they are all local parts of a global MULTISIM structure. Mismatch between referent and model systems behavior has been revealed in first runs, calling for modification of models. However, they have been implemented not only for the purpose of arriving at adequate descriptive and explanatory vehicles, but also to allow for exploration of some of the problems involved in a multiple-oriented approach. Two such problems concern the need for redefinition of levels with shift of perspective and for redefinition of time with shift of level.

Such problems were actualized through the evoked empirical and computer-operational procedures. This is consistent with our recurrent experience that neat paradigms for systems and theory articulation retain their neatness only as long as they are not confronted with execution of procedures of empirical-operational, computer-operational, or participant-operational kinds. That is why viewpoints of systems research as one-way, top-down postulational approaches need be replaced by two-way mediational viewpoints. The latter is reflected by Cavallo and Klir (1978).

The tendency toward formal closure and context insensitivity in general systems research, however, may seem to conflict with this mediational viewpoint. The generality of a systems paradigm need not imply a hierarchical layout. Problem sensitivity requires awareness of shifts in contexts. The principle of redundancy of potential commands (McCulloch, 1965) and storage in parallel of a multiplicity of correction-open elements seem to be called for in systems paradigms of relevance to psychosocial sciences. The repertoire of computer-, participant- and empirical-operational procedures offered by such paradigms need take into account the fact that psychosocial systems have the capacity of shifting between and transcending different sets of "laws" or regularities. This requires a theory in its own right. The present research merely touches at some fringe aspects of the problems involved.

REFERENCES

Abelson, R.P., E. Aronson, W.J. McGuire, T.M. Newcomb, M.J. Rosenberg, and P.H. Tannenbaum, eds. 1968. *Theories of Cognitive Consistency*. Chicago: Rand McNally.

Abelson, R.P., and A. Bernstein. 1963. "A Computer Simulation Model of Community Referendum Controversies." *Public Opinion Quarterly* 27:93–122.

Birthwistle, G., O.J. Dahl, B. Myrhaug, and K. Nygaard. 1973. *SIMULA Begin*. New York: Petrocelli Charter; Lund: Studentlitteratur.

Bråten, S. 1966. *Markedskommunikasjon*. Oslo: Inst. for Markedsføring. (Swedish edition Beckmanns, Stockholm, 1968).

——. 1968. "A Simulation Study of Personal and Mass Communication." *IAG Journal of IFIP Administrative Data Processing Group* 2:7–28.

——. 1971. "The Human Dyad." IMAS Information no. 24E0271, Solna. (Extended into Working Paper No. 72, University of Oslo, Institute of Sociology, 1977.)

——. 1972. "Simulering av Moralsk Dilemmabehandling Under Diadisk Samhandling." Bergen. (Translated into "Computer Simulation of Dilemma

Processing Dyads." Working Papers No. 73 a/b, University of Oslo, Institute of Sociology, 1977.)
——. 1973. "Kodingskretsløp Under Symbolsk Samhandling." *Tidsskr. for Samfunnsforskning* 14(2):47–63.
——. 1976. "En Konsistens og Kommunikasjonsmodell – Som Tillater Simulering av EF-striden." *Tidsskr. for Samfunnsforskning* 17(2):158–97.
——. 1977. "Computer Simulation of Consistency and Communication." In *VIIIth International Congress on Cybernetics. Proceedings.* Namur.
Bråten, S., and U. Norlén. 1973. "Simulation Model Analysis and Reduction." In W. Goldberg, ed. *Simulation vs. Analytical Solutions for Business and Economic Models. Proceedings,* vol. 1. BAS 17: Gothenburg University.
Cartwright, D., and F. Harary. 1965. "Structural Balance: A Generalization of Heider's Theory." *Psychological Review* 63:277–93.
Cavallo, R., and G.J. Klir. 1978. "A Problem-Solving Basis for General Systems Research." In G.J. Klir, ed. *Applied General Systems Research.* New York: Plenum Press.
Coleman, J., E. Katz, and H. Menzel. 1957. "The Diffusion of an Innovation among Physicians." *Sociometry* 20:253–70.
Dahl, O.J., and K. Nygaard. 1965. *SIMULA.* Oslo: Norwegian Computing Center.
Guetzkow, H., P. Kotler, and R. Schultz, eds. 1972. *Simulation in Social and Administrative Science.* Englewood Cliffs, N.J.: Prentice-Hall.
Jahren, E. 1973. "Analyse av et nettverkspanel i EF-saken." Bergen: NAVF's Sekr. for Mediaforskning. (Cand. Sociol. thesis, University of Oslo, 1973.)
Katz, E., and P.F. Lazarsfeld. 1955. *Personal Influence.* New York: Free Press.
Klir, G.J. 1969. *An Approach to General Systems Theory.* New York: Van Nostrand.
——. 1975. "On the Representation of Activity Arrays." *International Journal of General Systems* 2(3):149–68.
Klir, G.J., and M. Valach. 1967. *Cybernetic Modelling.* London: Iliffe Books.
McCulloch, W.S. 1965. *Embodiments of Mind.* Cambridge, Mass.: MIT Press.
Merton, R.K. 1975. "Structural Analysis in Sociology." In P.M. Blau, ed. *Approaches to the Study of Social Structure.* New York: Free Press.
Mesarovic, M.D., D. Macko, and Y. Takahara. 1970. *Theory of Hierarchical Multilevel Systems.* New York: Academic Press.
Næss, A. 1972. *The Pluralist and Possibilist Aspect of Scientific Enterprise.* Oslo: Universitetsforlaget; London: Allen & Unwin.
Norlén, U. 1972. "Simulation Model Building." Ph.D. dissertation, Gothenburg University, Department of Statistics. (Halsted Press, Wiley, 1975.)
Rogers, E., and L. Svenning. 1969. *Modernization among Peasants.* New York: Holt, Rinehart, & Winston.
Zeigler, B.P. 1976. *Theory of Modeling and Simulation.* New York: Wiley.

8 HYPOTHESES BEHIND THE SOCIOLOGICAL INTERVIEW: *Test and Reformulation*

Johannes van der Zouwen

FORMULATION OF A CONCEPTUAL MODEL OF THE INTERVIEW PROCESS

The interview is an indispensable method for gathering information concerning opinions, motives, and attitudes of human actors. Like every research method, the interview rests on a number of auxiliary hypotheses. For the description and classification of these hypotheses, a conceptual model of the interview was developed; it was based on an analogy with black box analysis.

A number of auxiliary hypotheses were inferred from common interview practice. From the conjunctions of these hypotheses we derived testable predictions. These were then confronted with data from a number of investigations concerning the interview. The predictions often were not borne out if the interview questions concerned opinions and attitudes about disputable affairs. In these cases, the probability is high that wrong conclusions are inferred from interview data.

These findings underlie the need for a correction of the auxiliary hypotheses (and as a consequence) of the instructions used for the interview process. Problems with the reformulation of the model of the interview process are presented in this chapter.

Many theories in the social sciences contain propositions in which a (causal) relation is assumed between opinions, preferences, or attitudes of people on the one hand, and their behavior on the other. We also encounter propositions in which a (causal) relation is assumed between a person's occupation or level of education on the one hand and his opinions, preferences, or attitudes on the other. Both types of propositions contain variables that are not directly observable for a researcher, such as opinions and attitudes. An analogy here with the black box is quite suggestive (van der Zouwen, 1974).

In the first type of proposition mentioned (Type A), a relation is assumed between an unobservable state variable (x) and an observable output variable (y), which is (partly) dependent on the state variable. Stated in other terms, an output function (r) is here postulated $y = r(x)$.

The propositions of the other type (Type B) contain an assumed relation between an observable input variable (u) and an unobservable state variable. Type B propositions, therefore, postulate a certain state transition function (s): $x = s(u)$. Figure 8.1 shows both types of sociological/psychological propositions stated in black box terminology.

To test these propositions, the researcher is obliged to use auxiliary hypotheses that lead to an estimation of the unobservable variable x. The researcher attaches an additional information channel to the black box for specific input and output signals and then postulates functions that relate these inputs and outputs to the state variable of the black box in which he or she is interested.

The interview is an example of this approach: by means of a specific additional input, a question (u_q), one receives an additional output, the answer of the respondent (y_a). The researcher assumes that a certain relationship exists between the answer the respondent gives and the state variable x referred to in question u_q. In other words, the researcher assumes that the respondent "used" a specific output function in answering u_q, namely response function r_a. With the help of an estimation of this response function ($\hat{r}_a$), it is possible to make an estimation of x ($\hat{x}$) on the basis of the answer the respondent gives (Dijkstra and van der Zouwen, 1977). (See Figure 8.2.)

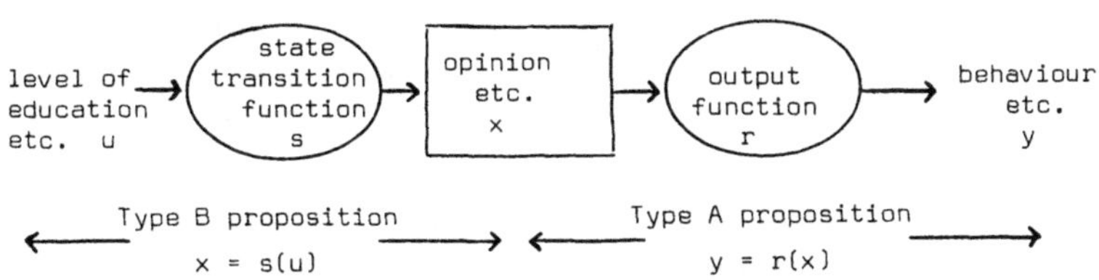

Figure 8.1. Type A and B Propositions in Sociological/Psychological Theories

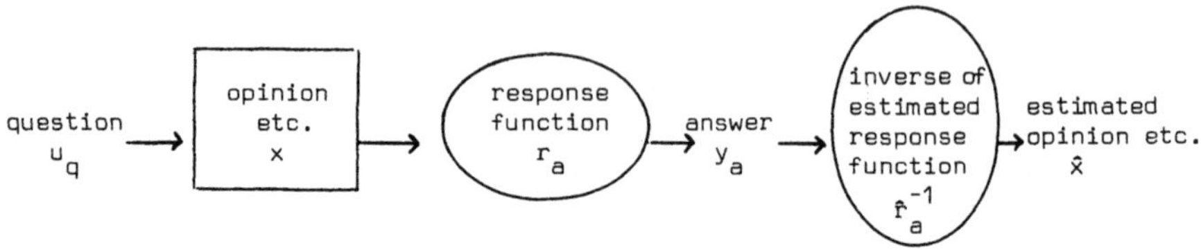

Figure 8.2. The Assumed Relation between Answer and Opinion

Now questions are not posed "in a vacuum" but in the context of a specific questionnaire by a specific interviewer in a specific interview situation. These characteristics of the interview are described with variables $u_1{}^*, u_2{}^*, \ldots$, etc., in short with vector $\mathbf{u}^*$. The researcher is obliged to make assumptions concerning the relationships between question u_q and interview characteristics $\mathbf{u}^*$ on the one hand, and state variable x on the other hand. That is, the researcher has to assume a state transition function s_q, relating u_q and $\mathbf{u}^*$ with x. (See Figure 8.3.)

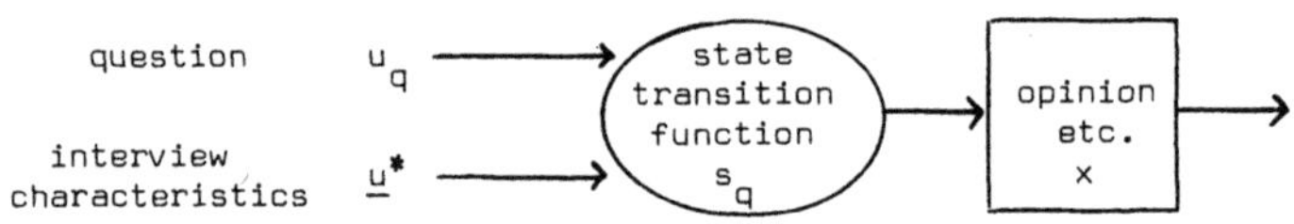

Figure 8.3. The Relationship between u_q, $\mathbf{u}^*$ and x

INFERENCE OF AUXILIARY HYPOTHESES BEHIND THE SURVEY INTERVIEW

The researcher, by inferring the "true" opinion of the respondent from his answer to the interview question, has to use two types of auxiliary hypotheses. Type 1 concerns the response function r_a; type 2 concerns the state transition function s_q. These auxiliary hypotheses concerning the interview may differ greatly, but they have at least one thing in common in that they all refer to a relationship between an observable (y_a, u_q, $\mathbf{u}^*$) and an unobservable quantity (x). The implication is that it is impossible to empirically test individual auxiliary hypotheses. This state of affairs is depicted in Figure 8.4, a combination of the foregoing figures, where unobservable quantities and functions are denoted by italics.

It is possible, however, to derive testable predictions from conjunctions of auxiliary hypotheses. The outcome of the test of such a prediction may give an indication concerning the tenability of the hypotheses from which the prediction is derived. We shall illustrate this procedure for the following auxiliary

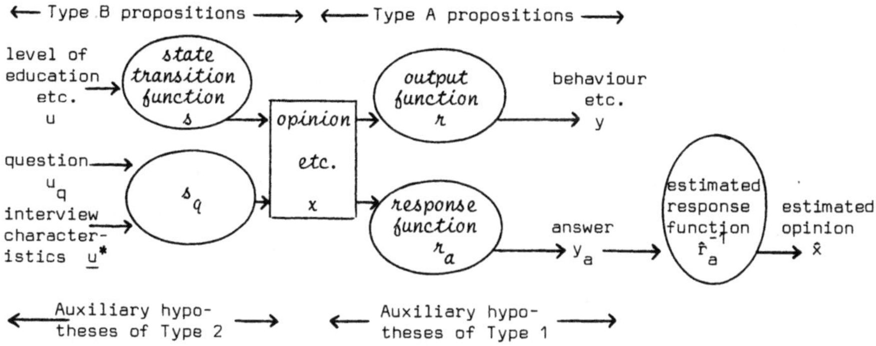

Figure 8.4. Observable and Unobservable Quantities in the Interview Process

hypotheses, apparently assumed by researchers who use a very popular type of interview, namely the survey interview:

Auxiliary hypothesis 1: When, as is common in survey research, the researcher uses the same coding scheme for every interview protocol, he apparently assumes that *all (n) respondents used the same response function* in answering a specific question.

Auxiliary hypothesis 2: Furthermore, the researcher usually assumes that the respondent understands the question posed in the same way as was meant by the researcher, namely as concerning only variable x. In other words, the researcher assumes that *the answer y_a is only dependent on the response function r_a used and variable x.*

Auxiliary hypothesis 3: In survey research one does not interview the total research population but only a representative sample drawn from this population. From the sample data the researcher makes inferences for the whole population. This procedure only makes sense if one assumes that *being interviewed as such has no effect on the opinion of the person interviewed.*

Auxiliary hypothesis 4: Since interview protocols contain very little information concerning the person of the interviewer or the situation in which the interview is held, one can derive that researchers assume that *so-called u^*-variables are not related with x.*

Auxiliary hypothesis 5: In common survey interviews the interviewer is obliged to adhere to the formulation and sequence of the questions as printed in the questionnaire. This prescription appears to be based on the auxiliary hypothesis that states that *differences in formulation or rank order of the questions lead to differences in response functions* used by the respondents. It is also conceivable that the researcher believes there is only one possible (or best) formulation and place of the question, namely the one chosen in the questionnaire, whereby the auxiliary hypotheses mentioned above are valid.

Applying the general state transition function s to an interview, held between time t and $t + \Delta t$, where the input vector $\mathbf{u}$ consists of the components u_q and $\mathbf{u}^*$, we find

$$x(t + \Delta t) = s_q\,[x(t), u_q, \mathbf{u}^*]\,.$$

According to auxiliary hypotheses 3 and 4, the researcher apparently assumes that neither input-component influences the state variable x. More precisely, he assumes that the function relating $x(t + \Delta t)$ with $x(t)$ is such that

$$x(t + \Delta t) = s_q\,[x(t)] = x(t). \tag{8.1}$$

In addition, it follows that as far as stochastic u^* variables are concerned, measured at interval or ratio level, and linearly related with x, the researcher assumes that

$$\text{cov}\,(x, u_1^*) = \text{cov}\,(x, u_2^*) = \cdots = 0. \tag{8.2}$$

Taking auxiliary hypothesis 2 into account, it appears that the researcher assumes the output obtained during the interview (y_a) depends only on x. That is, the general output function may be written as

$$y_a(t + \Delta t) = r_a\,[x(t + \Delta t)]\,. \tag{8.3}$$

Moreover, it follows from auxiliary hypothesis 1, that for a given question u_q, the output function r_a is the same for all (n) respondents. That is,

$$r_{a1} = r_{a2} = \cdots = r_{an}\,. \tag{8.4}$$

Finally, auxiliary hypothesis 5 states that the researcher assumes that if there are m different formulations of question u_q concerning state variable x, to each formulation j there belongs an output function r_{aj} such that

$$r_{a1} \neq r_{a2} \neq \cdots \neq r_{am}\,. \tag{8.5}$$

DERIVATION OF TESTABLE PREDICTIONS

The conjunction of auxiliary hypotheses 1, 3, 4, and 5 leads to a prediction regarding the relationship between u_q formulations and the response. If u_q formulations differ from each other due to different values on (stochastic) formulation variables $f_1, f_2, \ldots, f_k$, and if these formulation variables are measured at interval or ratio level and are linearly related with y_a, this prediction reads

$$\text{cov}\,(f_1, y_a) \neq 0, \quad \text{cov}\,(f_2, y_a) \neq 0, \quad \text{and so on.} \tag{8.6}$$

We shall call this type of prediction a *covariance prediction.*

Another covariance prediction originates from the conjunction of the auxiliary hypotheses 2, 3, and 4, namely,

$$\text{cov}\,(y_a, u_1^*) = \text{cov}\,(y_a, u_2^*) = \cdots\cdot = 0. \tag{8.7}$$

Conjunction of hypotheses 1, 3, and 4 leads to a so-called *stability prediction,* namely, if

$$x(t + 1) = x(t),$$

then

$$y_a(t + 1) = y_a(t). \tag{8.8}$$

Even if the researcher were to fix the values on the formulation variables f by means of strict prescriptions for the interviewer, and if hypotheses 1 to 4 all were to hold for his research project, the researcher could still receive a wrong image of the opinions of his respondents. This has to do with the fact that the researcher, wanting to make an estimation of x, $(\hat{x})$, is forced to make an estimation of r_a, $(\hat{r}_a)$. Figure 8.2 shows how the researcher (after assuming that r_a is both one-to-one and onto) uses the inverse function of r_a, $(\hat{r}_a^{-1})$, to infer $\hat{x}$ from the answer the respondent gives.

The researcher usually assumes that his estimation of r_a is correct. Stated in other terms, this auxiliary hypothesis reads

$$\hat{r}_a = r_a. \tag{8.9}$$

If the researcher has at his disposal an estimation of x, $(\dot{x})$, that is assumed to be correct, $(\dot{x} = x)$, for example, administrative data regarding the question under concern, then it is possible to derive from (8.9) the prediction that if

$$\hat{r}_a = r_a \text{ and } \dot{x} = x,$$

then

$$\hat{x} = \dot{x}. \tag{8.10}$$

We call this prediction an *accuracy prediction.*

We have thus derived, from some current auxiliary hypotheses behind the interview, three types of testable predictions. (See Figure 8.5.)

Auxiliary hypotheses verbalized		Auxiliary hypotheses formalized		Testable predictions derived.	
a) All respondents use the same response function.	⟷	$r_{a1} = r_{a2} = \dots = r_{an}$	(4)	if $x(t+1) = x(t)$ then $y_a(t+1) = y_a(t)$	(8)
b) The answer of a respondent on an interview question is only dependent on the response function used and the opinion, etc. of the respondent	⟷	$y_a(t + \Delta t) = r_a(x(t + \Delta t))$	(3)		
c) Being interviewed as such has no effect on the opinion etc. of the person interviewed.	⟷	$x(t + \Delta t) = s_q(x(t)) = x(t)$	(1)	$\text{cov}\,(y_a, u_1^*) = \text{cov}\,(y_a, u_2^*) = \dots = 0$	(7)
d) Characteristics of the interview are not related to the opinion etc. of the respondent.		$\text{cov}(x, u_1^*) = \text{cov}(x, u_2^*) = \dots = 0$	(2)		
e) Differences in formulations or rank order of the questions lead to differences in response functions used by the respondents.	⟷	$r_{a1} \neq r_{a2} \neq \dots \neq r_{am}$	(5)	$\text{cov}(f_1, y_a) \neq 0$, $\text{cov}(f_2, y_a) \neq 0$, etc.	(6)
-) The estimation of the response function is correct.	⟷	$\hat{r}_a = r_a$	(9)	if $\hat{r}_a = r_a$ and $\dot{x} = x$ then $\hat{x} = \dot{x}$	(10)

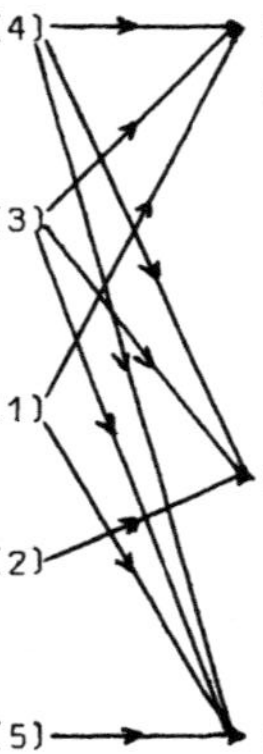

Legend:

- r_a : answer function of the respondent
- n : number of respondents in sample
- y_a : answer of the respondent
- t, $t + \Delta t$: time period in which interview is held
- x : opinion, etc. of the respondent
- s_q : state transition function relating u_q and u^* with x
- $u_1^*, u_2^*, \dots$: variables characterizing the interview (situation)
- m : number of different possible formulations of u_q
- $\hat{r}_a$: response function estimated by researcher
- $f_1, f_2 \dots$: variables characterizing different formulations of u_q
- $\dot{x}$: external information about x that is regarded as rather correct.
- $\hat{x}$: estimation of x on the basis of $\hat{r}_a$ and y_a

Figure 8.5. The Formalization of the Auxiliary Hypotheses and the Derivation of Testable Predictions

TEST OF THE PREDICTIONS

Comparison of Data from Different Projects

To confront the above-mentioned predictions with data from relevant research projects is to encounter the problem that some of these projects are hardly comparable. The more than hundred relevant projects differ not only as to their research topic, population, chosen method of interviewing, and so forth, but also as to their research designs. These projects, therefore, must first be classified according to their research design.

In comparing data from comparable projects, it appeared that researchers found that these results differ considerably: some outcomes support a prediction while other outcomes are incompatible with the same prediction. As a case in point, we mention that Athey et al. (1960) found that the race of the interviewer had considerable impact on answers concerning racial issues—contrary to prediction (8.2)—while Bryant et al. (1966) in a replication study did not find this relationship between race of the interviewer and answers given by the respondents. However, such contradictory research outcomes become understandable and manageable if one conceives research outcomes as drawings from a probability distribution. According to Taveggia (1974), research results are probabilistic: ". . . if a large enough number of researches has been done on a particular topic, chance alone dictates that studies will exist that report inconsistent and contradictory findings" (p. 398). So the advised solution is simply to accumulate research findings and examine their distribution.

In examining distributions of interview results, however, exploratory results often are concerned, and consequently, zero results, in contrast to positive or negative results, tend to be underreported. In addition, Taveggia's procedure is only useful if both positive and negative results are possible. The implication is that we need unidimensional variables, measured on at least ordinal level (or on nominal level with no more than two categories).

Formulation of More Flexible Predictions

The auxiliary hypotheses 1 through 5 were very strictly formulated: r_a is the same for *all* respondents, the covariance is exactly zero, there is no difference between $\hat{r}_a$ and r_a, and so on. In practice, the researcher will think in terms of "about the same," "differs insignificantly from zero," "the covariance is statistically insignificant," and so forth.

Less strict hypotheses lead to less strict predictions. To be able to test these less strict predictions, it is necessary to formulate measures for agreement and association, and one must indicate at which value of these measures a difference or association not worth mentioning becomes worth mentioning. For the analysis of the research data, we used current statistical measures (such as the product-moment correlation coefficient) and current criteria (such as the famous 5 percent significance level). We did so because we think that testing these predictions makes more sense if the auxiliary hypotheses from which these predictions are derived agree more with the intentions of the researcher who employs the interview as a method of data collection.

Missing Data

Not all testable predictions can be tested with the available data. For a correct test of a stability prediction $[y_a(t+1) = y_a(t)]$, one needs research data in which it is reasonable to assume that in the period between two subsequent interviews the state variable concerned remained *unchanged* $[x(t+1) = x(t)]$. But panel interviews are most often used to measure *changes* in opinions and attitudes, for example, the effects of propaganda. So these data are of no use for the test of stability predictions.

For an appropriate test of an accuracy prediction $(\hat{x} = \dot{x})$, one has to assume that the external estimation of x, $(\dot{x})$, is correct. This assumption, however, only makes sense for factual data, such as age or number of children. So accuracy predictions can only be tested for factual data and not for what is called "attitudinal information"—that is, "data on attitudes, intentions, expectations, personality attributes, self-judgments, and any other matter about which there is, in principle, no objective external evidence against which to verify the response" (Sudman and Bradburn, 1974, p. 7).

RESULTS AND CONSEQUENCES

The results of the data analyses are summarized in Table 8.1. We see that the predictions derived from the auxiliary hypotheses behind the survey interview are borne out when the questions concern factual information that the respondent can answer without much difficulty—for example, when he does not need to consult his memory, or when social sanctions do not affect his giving an honest response. If other types of data are regarded, however, especially data concerning attitudes about disputable affairs, the predictions are often not borne out.

The probability that errors in the measurement of the variables y_a, u^*, $\dot{x}$, and f caused the rejection of some predictions appears low. The prediction $\hat{x} = \dot{x}$

Table 8.1. Results of the Test of the Predictions

Prediction	*Appears to Be Rejected If . . .*	*Appears to Be Borne Out If . . .*	*Conditioning Terms*
Accuracy prediction: $\hat{x} = \dot{x}$ (8.10) (only testable for factual information)	x concerns past events and/or has categories with different degree of social desirability and especially when	x concerns recent events and/or if the categories of x have approximately equal social desirability and especially when	
	respondents have a low social status	respondents have a high social status	Social status of respondents
Covariance prediction: cov $(y_a, u^*) = 0$ (8.7)	u^* is the value on x for the interviewer (as perceived by the respondent) and especially when	u^* is the degree of rapport between interviewer and respondent	
	x concerns attitudinal information and/or		Type of information factual vs. attitudinal
	respondents have a low social status		Social status of respondents
Covariance prediction: cov $(f, y_a) \neq 0$ (8.6)	f is the place of the question in the questionnaire	f is wording of the question and especially when	
		respondents have a high social status	Social status of respondents
Stability prediction: $y_a(t_3) = y_a(t_2)$ (8.8)	relevant data $[x(t_3) = x(t_2)]$ missing		

(8.10) is only tested for factual information, u^* often denoted simple variables as sex or race of the interviewer or place of the interview, and the prediction $\text{cov}(f, y_a) \neq 0$ (8.6) is rejected if f denotes "place of the question in the questionnaire," a variable that can be measured without error. Subsequent data analysis (Dijkstra, 1977*b*) indicated that the effect of (characteristics of) the interviewer on the answers of the respondent is greatest if the respondent is of low social status. The effect of different formulations of the same question, however, appears to be greater if respondents of high status are concerned.

The results have rather far-reaching consequences: the probability that wrong conclusions will be inferred from interview data is highest precisely in those cases in which the interview is most difficult to replace by another observation technique, namely in research concerning attitudinal data. To avoid these wrong conclusions, corrections of several auxiliary hypotheses and, as a consequence, a reformulation of several instructions used for the interview process are required. Our results emphasize the relevancy of the instruction that "the interviewer is obliged to adhere to the formulation of the questions as printed in the questionnaire." The data also suggest two new instructions:

1. If the sample is very heterogeneous, different coding schemes for the interview protocols have to be developed to account for the differences in response functions used by respondents belonging to different subcultures.
2. To avoid systematic interviewer effects, interviewers have to be assigned randomly to respondents.

The last section of this chapter sketches some problems with the reformulation of the model of the interview process.

PROBLEMS WITH THE REFORMULATION OF THE INTERVIEW MODEL

Conditions for an Adequate Interview Model

Considering that at least two parties are engaged in the interview process (the respondent and the interviewer-researcher), an interview model has to deal with (1) the mental processes leading to an answer from the respondent and (2) the effects of activities of the interviewer/researcher on these mental processes. A prerequisite for an adequate interview model is that the researcher can derive from the model instructions and guidelines for his (interviewer's) activities and infer estimations of the functions s_q and r_a, which represent the mental processes just mentioned. Figure 8.6 illustrates the close relationship between interview model and interview method.

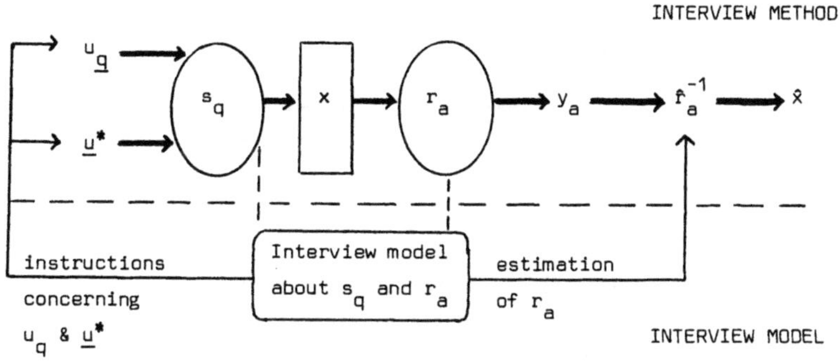

Figure 8.6. The Relationship between Interview Method and Interview Model

The researcher who wishes to make an estimation ($\hat{x}$) of the opinion (x) of a respondent on the basis of his answer (y_a) to an interview question needs answers to the following questions (see Figure 8.7):

1. Is y_a only dependent on x or are there other factors influencing y_a in a random (ϵ) or systematic (Z) manner?
2. Which factors (W) affect the choice of a certain response function (r_a) by a respondent?

Answers to both questions have to be included in a model of the interview process.

Nowakowska's Model as an Illustration of the "Testability Problem"

Several social scientists have published rather plausible verbal models of parts of the interview process—models that meet the conditions just formulated. Although formalizing these models and integrating them into a complete interview model would appear to be a good procedure, the difficulty is that, in relation to the number of observable variables, these models contain so many unobservable variables that the chances are small that they can be empirically tested. To illustrate this "testability problem," we present a model of the process of answering a questionnaire item as developed by Maria Nowakowska (1970). This researcher asked a jury to evaluate twenty-eight questionnaire items on seventeen 7-point scales. A factor analysis of the data showed seven "underlying" factors, of which the following five were included in her "answering model":

1. Emotional and motivational attitude;

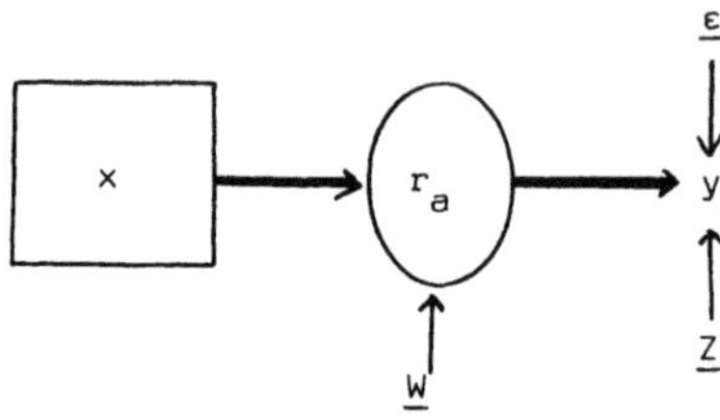

Figure 8.7. Factors Influencing the Answer of the Respondent

2. Specific (past) experience;
3. Intellectual evaluation of question and answer;
4. Value of question;
5. Social desirability.

These factors are interpreted as elements of a rather complicated information-transformation process that begins with the question and ends with the ultimate choice of one of the answer categories. Figure 8.8 represents Nowakowska's answering model, including some modifications we made to harmonize the verbal and pictorial representation of her model. In our opinion this sophisticated model seems plausible, but at the same time it appears to be completely untestable.

In order to construct a testable interview model that meets the conditions stated earlier, we are now developing a procedure for model construction with the following characteristics:

1. An inventory of auxiliary hypotheses related to the interview process is made, based on
   a. Existing interview models (Sudman and Bradburn, 1974; Getzels, 1954; Back, 1962; Back and Gergen, 1963; Kolson and Green, 1970; Williams, 1964; Summers, 1969; Nowakowska, 1970; Atteslander and Kneubühler, 1975);
   b. Instructions for interviewing and questionnaire construction as used in common interview practice and/or proposed in handbooks on interviewing (Kahn and Cannell, 1957; Gorden, 1975);
   c. Data from research concerning the interview (Dijkstra and van der Zouwen, 1977; Sudman and Bradburn, 1974);
   d. Social psychological theories concerning communication processes in dyadic relations between persons with very different roles (Dijkstra, 1977).

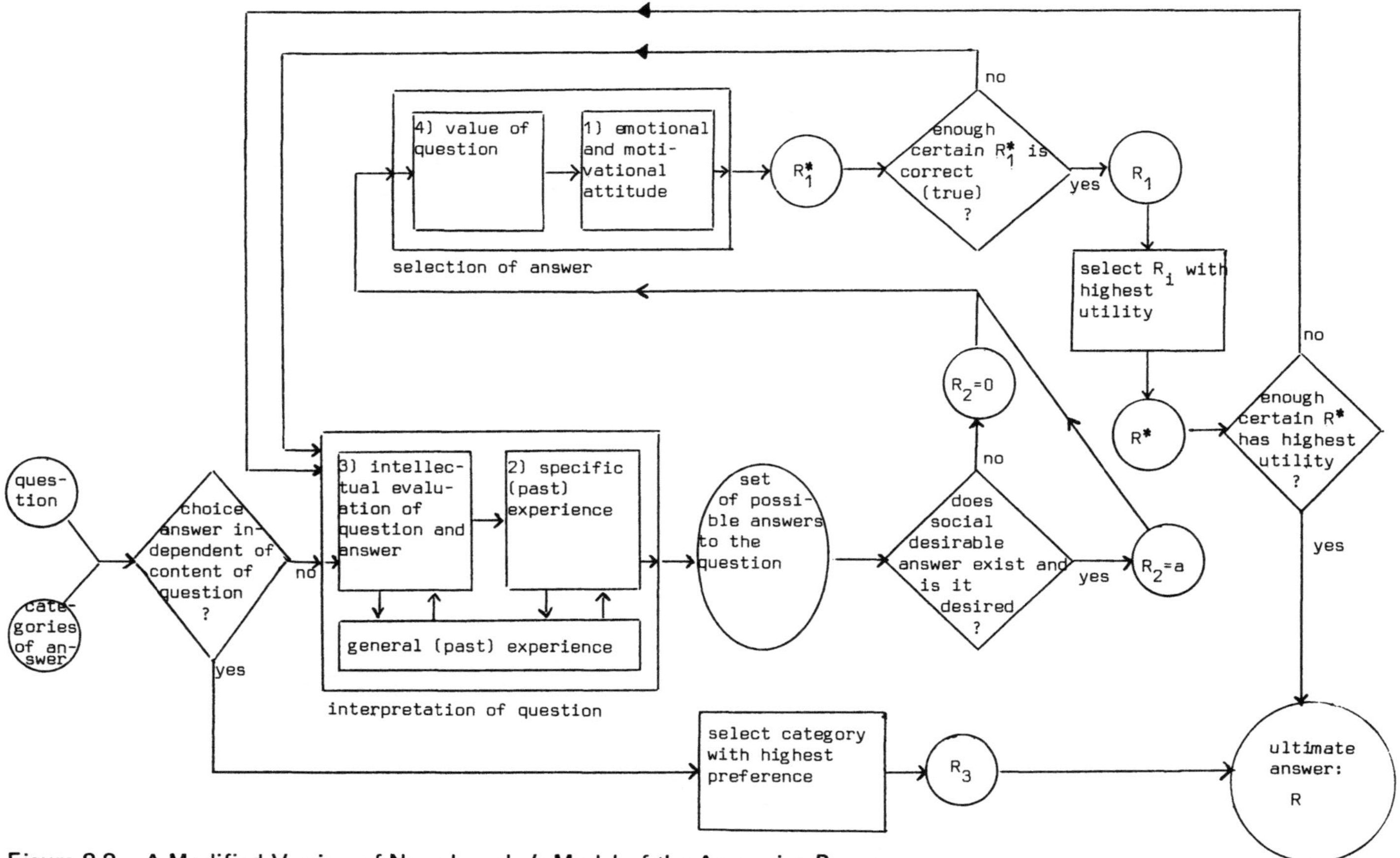

Figure 8.8. A Modified Version of Nowakowska's Model of the Answering Process

2. From conjunctions of these hypotheses, testable predictions are derived, which are then confronted with data from experiments especially designed for this purpose.
3. Hypotheses that stand this confrontation are integrated into an interview model that can be regarded as a specification of the proposed meta-model of the interview process (van der Zouwen, 1974). In this model the ratio between observable and unobservable variables must be high enough so that the parameters of the model can be estimated from (the structure of the model can be tested against) the data from the experiments mentioned above.

We hope to publish the results of this procedure within the foreseeable future.

REFERENCES

Athey, K.R., J.E. Coleman, A.P. Reitman, and J. Tang. 1960. "Two Experiments Showing the Effect of the Interviewer's Racial Background on Responses to Questionnaires concerning Racial Issues." *Journal of Applied Psychology* 44(4):244–46.

Atteslander, P., and H.U. Kneubühler. 1975. *Verzerrungen im Interview.* Opladen: Westdeutscher Verlag.

Back, K.W., 1962. "Social Research as a Communications System." *Social Forces* 41:61–68.

Back, K.W., and K.J. Gergen. 1963. "Idea Orientation and Ingratiation in the Interview: A Dynamic Model of Response Bias." *Proceedings of the Social Statistics Section of the American Statistical Association* 58:284–88.

Bryant, E.C., I. Gardner, and M. Goldman. 1966. "Responses on Racial Attitudes as Affected by Interviewers of Different Ethnic Groups." *Journal of Social Psychology* 70:95–100.

Dijkstra, W. 1977*a*. *Role Enactment in the Interview.* Free University, Amsterdam, Department of Research Methods.

——. 1977*b*. *Some Determinants of Interview-Bias.* Free University, Amsterdam, Department of Research Methods.

Dijkstra, W., and J. van der Zouwen. 1970. "Testing Auxiliary Hypotheses behind the Interview." *Annals of Systems Research* 6.

Getzels, J.W. 1954. "The Question-Answer Process: A Conceptualization and Some Derived Hypotheses for Empirical Examination." *Public Opinion Quarterly* 18:80–91.

Gorden, R.L. 1975. *Interviewing: Strategy, Techniques and Tactics,* rev. ed. Homewood, Ill.: Dorsey Press.

Kahn, R.L., and C.F. Cannell. 1957. *The Dynamics of Interviewing.* New York: Wiley.

Kolson, K.L., and J.J. Green. 1970. "Response Set Bias and Political Socialization Research." *Social Science Quarterly* 51:527–38.

Nowakowska, M. 1970. "A Model of Answering to a Questionnaire Item." *Acta Psychologica* 34:420–39.

Sudman, S., and N.M. Bradburn. 1974. *Response Effects in Surveys.* Chicago: Aldine.

Summers, G.F. 1969. "Toward a Paradigm for Respondent Bias in Survey Research." *Sociological Quarterly* 39:113–21.

Taveggia, T.C. 1974. "Resolving Research Controversy through Empirical Cumulation." *Sociological Methods and Research* 2(4):395–407.

Williams, J.A., Jr. 1964. "Interviewer-Respondent Interaction: A Study of Bias in the Information Interview." *Sociometry* 27:338–52.

Zouwen, J. van der. 1974. "A Conceptual Model for the Auxiliary Hypotheses behind the Interview." *Annals of Systems Research* 4:21–37.

9 A SYSTEM CONCEPT AND ITS IMPACT ON MULTIOBJECTIVE DECISION MODELS

Christer Carlsson

Most modelbuilders, and consequently model users, quite often find themselves in a dilemma. On the one hand, management normally involves decision making in all ill-structured, dynamic, and uncertain environment on the basis of rather messy conceptions of means and ends and interinfluences. On the other hand, most of the presently operational OR techniques are algorithms designed as aids for decision making and problem solving in a reasonably well-structured, rather certain, and scarcely dynamic environment. What, then, is to be done?

This chapter explores a systems concept that is applicable for an attack on the model user's dilemma. This concept is based on three central elements: *activity units, intrarelations,* and *interrelations,* which by definition form a set with a hierarchical structure. It is suggested that this systems concept could be applied for modeling decision-making situations of a dynamic and complex structure in which the central elements even could be of different levels of aggregation. The concept also seems to provide a necessary conceptual framework for a simultaneous implementation of a set of objectives, for coordinating and/or aggregating the objectives, and for constructing an algorithm, a program, a rule, and so forth, that would form necessary and/or sufficient conditions for an attainment of all the goals—either simultaneously, or each one of them, alternatingly—at least once in some chosen period of time.

"Problem" is one of those concepts we use in a catchall fashion in our everyday language; even among those regarded as "specialists" (operations researchers, systems theoreticians, mathematicians and behavioral scientists), a "problem" would refer to different conceptual units. Here we will deal with problems, and as we intend to explore the properties of a systems concept as a means for problem solving, we will have to formulate "problem" so that it is precise enough to expose the properties of the systems concept and sufficiently general enough to give possible results generality.

Management science deals with problems in a management context—problems described as messy, irregular, and frustrating—because the decisionmakers know that something should be done. However, they do not know what, how, and when—or sometimes even why. The traditional way to deal with problems is to simplify them in a model, obtain a solution in that model, and apply it to the "real" problem. But here a dilemma arises. As management normally involves decision making in an ill-structured, dynamic, and uncertain environment on the basis of more than one objective, most of the presently operational OR-techniques are aids designed for decision making and problem solving with one objective in a reasonably well-structured, rather certain, and scarcely dynamic environment.

We will try to find a way out of this dilemma through a systems concept we claim to be well suited for capturing essential features of decision problems created by nonstructure, a multiplicity of objectives, uncertainty, and dynamics. We will then introduce a special class of problems, *W*-problems, and find out if they can be made to represent the true complexity of a genuine management problem. Next we will introduce a systems concept and discuss its properties as a basis for a modeling and problem-solving technique. Finally, we will operationalize the systems concept.

COMPLEX PROBLEMS INVOLVING A SET OF OBJECTIVES

To outline what we mean by the "true complexity of a genuine management problem," let us assume that a decisionmaker (DM) is able:

1. To identify and formulate a set G of objectives that should be attained either all simultaneously at t_1, or each one of them at least once in an interval $[t_1, t_2]$;
2. To identify, with some certainty, a subset C of a set A of available activities, which the DM has found to be relevant for an attainment of G;
3. To delimit a set O, which represents the outcomes of applying the activities of A in an environment E, and at least identify and formulate elements of O and E that are relevant for C;
4. To give some hints on a set M of relations existing between C, E and O.

Let ⟨*A, C, E, G, M, O*⟩ be DM's decision-context, which is assumed to be a collection of all available information, and let us further assume that the DM finds it complex in the following sense (see Figure 9.1):

1. The elements of *A* – the DM's available activities – support or counteract each other in various combinations and in various ways, of which DM, however, does not have any definite information.
2. An activity produces some effects – the elements of *O* – which are either fully known, partially known, or completely unknown to DM.
3. An element of *E* influences, and is influenced by, an activity, a subset of activities, or by the whole set *A* (DM may not have sufficient information on the actual forms for interaction).
4. Effects generated by an activity are at least partly determined by one or some elements of *E*.
5. An objective – an element of *G* – is defined either for an activity, some subset of activities, or for the whole set *A* (an objective can also, correspondingly, be defined for the set *O*).

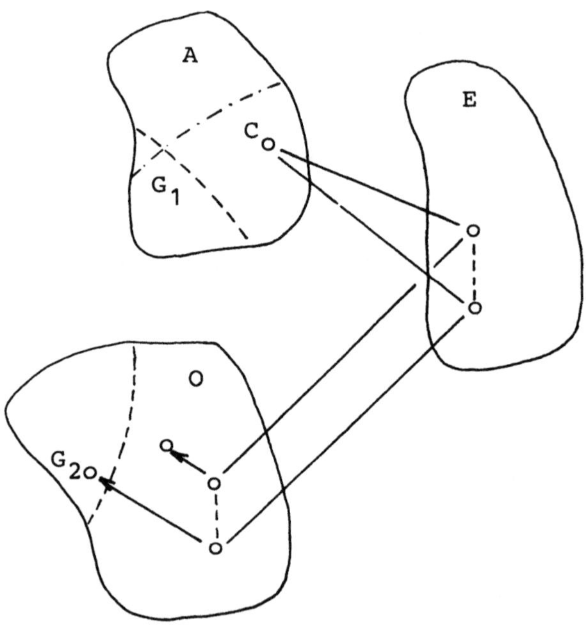

Figure 9.1. An *M*-Situation

This means, in the present formulation (see Figure 9.1), that DM's choosing to apply an activity (from C) will produce some consequences: (1) other activities he may consider in the next steps will be affected; (2) he might not be able to predict the effects (in O) of applying the activity; (3) he has some information on how a chosen environment (E) interacts with the chosen activity; and (4) some of the effects represent an attainment of an objective (G_2) at some t_1 or at least once in an interval $[t_1, t_2]$. These characteristics represent, essentially, a fairly realistic context for a decisionmaker.

What we mean by the "true complexity of a genuine management problem" is that DM, under the above conditions, should either choose activities to attain the objectives all at once at some chosen t_1, or choose activities individually at least once in an interval $[t_1, t_2]$. Here our sense of "complexity" is influenced by May (1972), Newell et al. (1958), Kaimann (1974), and Sahal (1976).

Let us then assume that DM is able to formulate the following relation from his decision context:

$$G_1, G_2 \subset G; G_1 \cap G_2 = \emptyset \quad \text{for all } G_i; \tag{9.1}$$

any $G_i \notin A, O$ is disregarded as irrelevant for the discussion.

$$M \subset O \times C \times E \times O, \tag{9.2}$$

where M is formulated as a subset of the outcome.

$$C \subset A; \tag{9.3}$$

that is, all activities have some outcome, but only a subset C of A produces an outcome that is $\subset G_2$.

If DM then conceptualizes the unspecified sets C, E, and O to obtain their abstract counterparts—but is content with G_1 being $= \emptyset$—and correspondingly transforms (9.2) so that the subset M is defined by

$$M : O \times C \times E \rightarrow O, \tag{9.4}$$

and

$$(o_1, c_1, e_1, o_2) \in M \quad \text{and} \quad (o_1, c_1, e_1, o_3) \in M \quad \text{implies} \quad o_2 = o_3, \tag{9.5}$$

where o, c and e are elements of O, C and E. That is, M is a function mapping the set (or some subset of) $O \times C \times E$ into O. Then the relations (9.1) to (9.5) imply that DM is able to do the following: (1) distinguish between two different sets of objectives; (2) consider only those objectives that can be implemented for available activities or their effects; (3) identify relations between activities that are effective for an attainment of G, observed outcomes, and an observed environment; and, finally, (4) trace observed outcomes to relevant activities and some state of environment.

Given (9.4) and an element $c \in C$, we can then define two sets O_c and O_e as follows:

$$O_c = \{o \mid (e)(o') [(o,c,e,o') \in M]\}, \quad (9.6)$$

$$O_e = \{o \mid (c)(o') [(o,c,e,o') \in M]\}, \quad (9.7)$$

from which it follows that if $(o,c,e,o') \in M$, then $o \in O_c \cap O_e$. If DM then makes the assumption that if

$$o \in O_c \cap O_e, \quad \text{then} \quad (\exists o') [(o,c,e,o') \in M], \quad (9.8)$$

it follows that $\langle A, C, E, G, M, O\rangle$ is an M-situation (cf. Banerji, 1969; M-situations were formulated by Marino, W-problems by Windeknecht), which means that there exists some ready formalism that could be applied to our rather vague formulation of DM's decision-making situation.

An M-situation has (by definition) certain properties that can be used to formulate a decision-making strategy for DM. It has long been a traditional strategy for dealing with complex decision-making situations to rework them as precise decision problems in terms of one, or a few, explicit goals, relevant activities, and a defined environment. We will apply the same strategy and reformulate the M-situation as a W-problem (cf. Banerji, 1969).

DM's decision-making situation is called a problemlike M-situation if and only if

$$E = \{e'\}, \text{ a unit set,} \quad (9.9)$$

$$o \in O_c \quad \text{for } c \in C \quad \text{implies that} \quad o \notin O_d, \quad (9.10)$$

where $O_d \subset O$ and $G_2 \cap O_d = \emptyset$;

$$O_e = \emptyset, \text{ the empty set,} \quad (9.11)$$

which means that the M-situation is reduced to the situation shown in Figure 9.2. The reduction means that outcomes generated by the activities of C are identifiable and are the only outcomes to be considered in the decision problem.

Given a problemlike M-situation $R = \langle A, C, \{e\}, G_2, M, O\rangle$, we can define a set $P(R) = \langle O, F, Q\rangle$, where $Q = G_2$, and F is a set of relations defined by

$$f \in F \quad \text{if and only if there exists a } c \in C \text{ such that, for all } o, o' \in O, (o, o') \in f \text{ if and only if } (o, c, e', o') \in M. \quad (9.12)$$

That is, DM should be able to single out those activities that form proper responses to the outcomes, which, in turn, are generated through one or more relations.

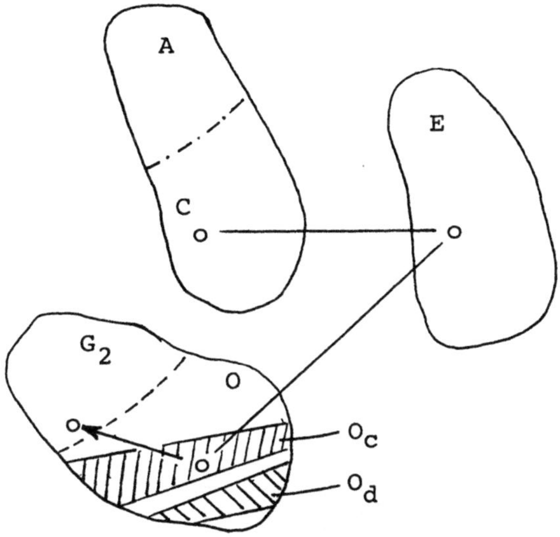

Figure 9.2. A Problemlike *M*-Situation

O is an abstract set and Q is a subset of O, and each element $f \in F$ is a function, since $(o, o') \in f$ and $(o, o'') \in f$ implies that $(o, c, e', o') \in M$ and $(o, c, e', o'') \in M$ for some unique c, so that $o' = o''$ by (9.5). Also $O_{fc} = o \in O_c$ implies that $o \in O_c \cap O_e$, whence by definition there exists an o' such that (o, c, e', o') $\in M$ whence f_c is defined for o proving $O_c \in O_{fc}$. Similarly, if $o \in O_{fc}$, then there is an o' such that $(o, c, e', o') \in M$, whence $o \in O_c$, proving the reverse inequality.

Thus there is a function $f_c \in F$ for each $c \in C$, and these are the only members of F. We have therefore made a further reduction and delimitation of a problemlike M-situation, as DM now is expected to formulate functions that could be used to link activities and outcomes in a way appropriate for an attainment of the elements of G_2 (see Figure 9.3). The set $P(R)$ is called a W-problem, and is thus a member of a class of problems with some interesting features for problem-solving and decision-making processes. We will concentrate on one of these features.

Given a W-problem and an element $o_j \in O$, a *winning solution* for o_i is a sequence of functions $f_1, f_2, \ldots, f_n$ such that $f_j \in F$ for each j, and such that $f_n(f_{n-1}(\ldots f_1(o) \ldots)) \in Q$. A function,

$$S: \bigcup_{f \in F} O_f - Q \rightarrow F \quad \text{is called a } W\text{-strategy if and only if } S(o) = f \text{ implies that}$$

$$o \in O_f. \tag{9.13}$$

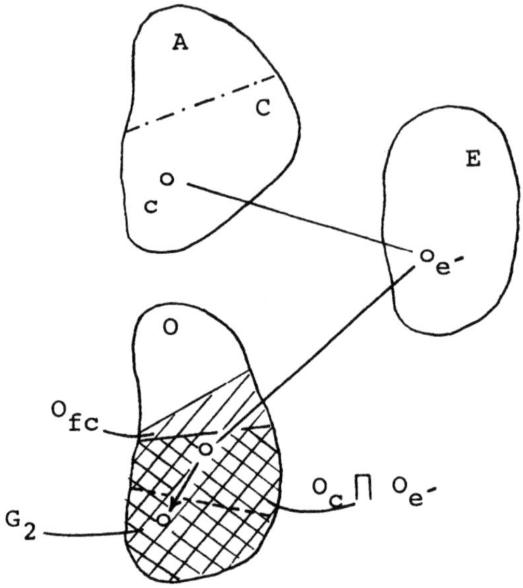

Figure 9.3. A *W*-Problem

A *W*-strategy is *winning* for $o_i \in O$ if there exists a winning solution $\{f_1, f_2, \ldots, f_n\}$ such that $f_1 = S(o_i)$, and for each j $(1 \leqslant j \leqslant n)$ $f_{j+1} = S(f_j(f_{j-1}(\ldots f_1(o_i) \ldots)))$. Another feature of *W*-problems (P) and problemlike *M*-situations (R) is that $P[R(P)] = R$, and $R[P(R)] = R$. A function,

$$P\colon O - (G_2 \cup O_d \cup O_e) \to C \text{ is called a } \textit{control strategy} \text{ if } P(o) = c \text{ implies } o \in O_c. \tag{9.14}$$

Given an element $o_i \in G_2 \cup O_d \cup O_e$, a control strategy P_C is called a *winning strategy* for o_i (see Figure 9.4) if there exists an integer N such that for any influence from the environment there exists a sequence $(c_1, e_1), (c_2, e_2), (c_3, e_3), \ldots, (c_n, e_n)$ such that $n \leqslant N$ and $c_1 = P_C(o_i)$, $e_1 = P_E(o_i)$, where P_E is the influence from E and for each $j (1 \leqslant j \leqslant n)$,

$$c_{j+1} = P_C((c_j, e_j)((c_{j-1}, e_{j-1})(\ldots(c_1, e_1)(o_i)\ldots))),$$
$$e_{j+1} = P_E((c_j, e_j)((c_{j-1}, e_{j-1})(\ldots(c_1, e_1)(o_i)\ldots))),$$

and

$$(c_n, e_n)((c_{n-1}, e_{n-1}))\ldots(c_1, e_1)(o_i)\ldots) \in G_2. \tag{9.15}$$

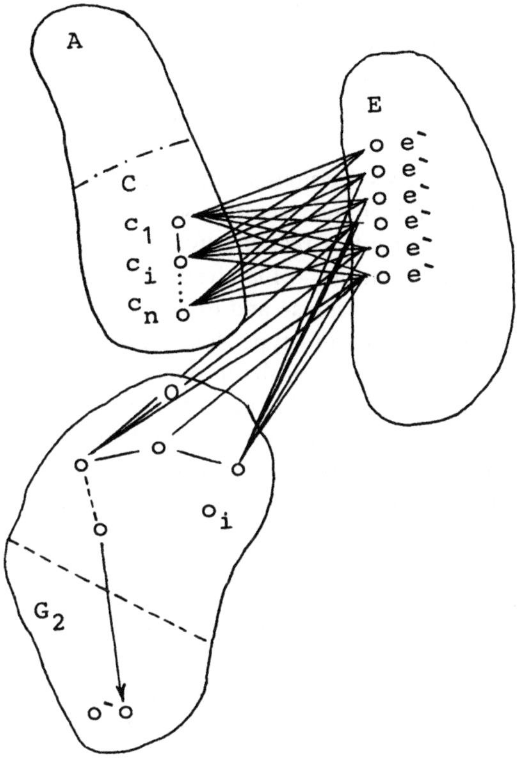

Figure 9.4. A Winning Strategy for a *W*-Problem

Let S be a *W*-strategy for the *W*-problem P. Let $R(S)$ be the control strategy for $R(P)$. Then S is a winning *W*-strategy for o_i if, and only if, $R(S)$ is a winning control strategy for o_i in $R(P)$. (9.16)

Banerji gives a proof for (9.16), but we will instead outline its implications in terms of DM's decision-making and problem-solving behavior:

1. If DM's decision-making situation can be formulated as a problemlike *M*-situation and further reduced to a *W*-problem, and
2. If DM knows, or could find out, the outcomes resulting from a series of activities ($\in C$) in a known environment E, and reduced to and described by a unit set $\{e^{`}\}$, and
3. If DM can formulate and implement a finite sequence of mapping functions—$O \times A \times E \rightarrow O$—which show how a certain outcome could be realized,

4. Then it would be possible for DM to describe how some wanted outcome should be realized—and one or more objective at least partly attained; this would solve the W-problem and also give DM precise directions for his problem-solving and decision-making process.

The transformation of DM's decision-making situation to an M-situation and a W-problem follows traditional management science and operations research methodology; it is an exponent of *reductionism*—an aspiration to reduce complex wholes to manageable entities—of *analysis*—a search for a few and simple explanatory cause-effect relationships—and of *mechanization*, as we are looking for explanatory relationships of the type "one cause, one effect." This methodology has won some remarkable results in operations research, but is it really relevant for dealing with complex problems involving more than one objective?

In a W-problem, there are no assumptions on interactions between the activities of A, and possible effects of such interactions are disregarded. The elements of O, or at least an active subset of O, are assumed to be fully known to DM. The environment E is represented by a unit set $\{\grave{e}\}$, and simple cause-effect relationships exist and can be implemented as mapping functions for the sets A(or C), E and O. This means that most of the complexity of a "genuine management problem" has been eliminated in the process of constructing a W-problem (although that is a relatively complete formulation).

Then, what about the case of multiple objectives? In (9.1) we assumed that it was possible to split DM's set of objectives G into two disjoint subsets G_1 and G_2, which were implemented for A and O respectively. But in the process of formulating a W-problem, we assumed that $G_1 = \emptyset$ and G_2 a subset of O, with no provisions for the difficulties involved in implementing more than one objective. Thus also, the multiobjectivity problem has been eliminated.

Is (9.16) then an appropriate decision model for DM—or, in other words, is (9.16) an appropriate decision model for dealing with genuine management problems? It eliminates two crucial aspects on management problems, complexity and multiobjectivity, and is thus to be regarded as less satisfactory. And not surprisingly, this is a common denominator for all recent criticism of operations research models and methodology (see Ackoff, 1974).

A SYSTEMS CONCEPT AND COMPLEX, MULTIOBJECTIVE DECISION PROBLEMS

In an attempt to find a way to deal with complex and multiobjective decision problems, we will introduce a systems concept, outline its properties in some detail, and then point out similarities to and differences from the formulation of the W-problem. This information will be used, finally, for outlining its impact on a methodology for developing and implementing decision models reflecting "the true complexity of genuine management problems."

Let us assume that A, the set of activities introduced earlier, is a finite set of relevant descriptions of the activities DM finds appropriate for an attainment of the set of objectives G in the environment E. We will furthermore assume that A is conceptually described as follows:

The set of activities, A, consists of two parts, or subsets, A_1 and A_2. (9.17)

These, in turn, are described conceptually by the following:

A_1 has a finite number of elements and is organized with an implemented ordering rule (such as $>, \geqslant, <, \neq$, etc.). (9.18)

Let $A_k^{(h)}$ be an element of A_1. It is a finite set of activities $a_{ki}^{(h)}$, each of which is represented by its numerical description $\alpha_{ki}^{(h)}$ [thus $A_k^{(h)} = \cup\ \alpha_{ki}^{(h)}$]; $\alpha_{ki}^{(h)}$ refers to a point of time t_1, or an interval $[t_1, t_2]$; the index h ($\in H$, a finite, numerical set that is organized according to the ordering rule) is an implementation of the ordering rule for an activity $a_{ki}^{(h)}$ or a set of activities $A_k^{(h)}$. (9.18*a*)

Let $a_{k1}^{(h)}$ and $a_{k2}^{(h)}$ be two activities ($\in A_k^{(h)}$); if they appear in some sense "together" at t_1, or in $[t_1, t_2]$, this is indicated with a relation that "joins" the two activities. (9.18*b*)

Let B be a set of relations with a finite number of elements; let each of its elements $B_k^{(h)}$ be an element of A_1 and represented by a numerical description ($\beta_{k12}^{(h)}$ for the activities $a_{k1}^{(h)}, a_{k2}^{(h)}$; $\{\beta_k^{(h)}\}$ for $A_k^{(h)}$; etc.); $B_k^{(h)}$ refers to a point of time t_1 or an interval $[t_1, t_2]$; the index h is an implementation of the ordering rule. (9.18*c*)

A_2 has a finite number of elements, and is organized with an implemented ordering rule (see 9.18). (9.19)

Equivalent with (9.18*a*); A_2 consists, consequently, of the same activities as A_1. (9.19*a*)

Consider three activities $a_{k1}^{(h)}$ ($\in A_k^{(h)}$) and $a_{k2}^{(\bar{h})}, a_{k3}^{(\bar{h})}$ ($\in A_k^{(\bar{h})}$), where $h > \bar{h}$; if $a_{k1}^{(h)}$ "consists of" $a_{k2}^{(\bar{h})}$ and $a_{k3}^{(\bar{h})}$ at t_1, or for an interval $[t_1, t_2]$, this is indicated with a relation that "joins" the two activities. (9.19*b*)

Let C be a set of relations with a finite number of elements; let each of its elements C_m be an element of A_2 and represented by a numerical description ($\gamma_{m123}^{(h)}$ for the activities $a_{k1}^{(h)}$, $a_{k2}^{(\bar{h})}$, $a_{k3}^{(\bar{h})}$; $\{\gamma\}_m^{(h)}$ or C_m for $A_{k1}^{(h_1)}$, $A_{k2}^{(h_2)}$, $A_{k3}^{(h_3)}$, where $h_1 > h_2 \geqslant h_3$;etc.); C_m refers to a point of time t_1 or an interval $[t_1, t_2]$; the index h (in $\gamma^{(h)}$) is an implementation of the ordering rule. (9.19*c*)

A is formed as some proper combination of A_1 and A_2; if A_1 and A_2 are appropriate numerical sets, this combination could be obtained as a Cartesian product. (9.20)

Following Mesarovic (1972), "A system is (or reflects the existence of) a relationship between the objects [of study]," we will furthermore assume that the relations introduced through (9.18*b* and 9.18*c*) and (9.19*b* and 9.19*c*) represent dynamic relationships, which we will generalize into two functional categories:

The relation "consists of" (see 9.19*b*) is a *transformation* of corresponding numerical descriptions; the relation "together" (see 9.18*b*) is a *coordination* (without transformation) of corresponding numerical descriptions (all the sets involved are assumed to be numerical). (9.21)

Let us now assume that the ordering rule ">" has been applied in such a fashion that H has a hierarchical structure; then it is possible to relate (9.21) to its structure (see 9.18*b* and 9.19*b*): a transformation is assumed to be a "vertical" operation, and a coordination correspondingly of a "horizontal" operation.

Horizontal relations, which we will call *intrarelations,* were represented by $\beta_{kij}^{(h)}$ and will, in order to simplify the discussion, also be taken to represent the *state* of the intrarelation at some t_1, or in an interval $[t_1, t_2]$; then an O-state means that the intrarelation is inactive or missing. Correspondingly, we will call vertical relations *interrelations*; $\gamma_{mijk}^{(h)}$ was taken to represent an interrelation, and we will also assume that it represents the *state* of that interrelation at some t_1, or in an interval $[t_1, t_2]$; an inactive or missing interrelation would then be represented by an O-state.

As the set H was given a hierarchical structure (see above), we can let the indices h represent *hierarchical* levels, and hence an intrarelation would coordinate activities of the same hierarchical level, but an interrelation would carry out transformations over different hierarchical levels.

A hierarchy is usually created by an asymmetric relation of domination or subordination (see Bunge, 1969), but as shown by Mesarovic et al. (1969), a system with different levels of description or abstraction would also have a hierarchical structure. This suggests that if the interrelations are appropriately specified, the structure of the set A could be linked with the hierarchical structure of H. We will suggest that the transformation carried out by the interrelations is one of *aggregation/disaggregation*—the former a result of moving "upward" in the hierarchy, the latter of moving "downward."

The set A of Figure 9.5 is our initial set of activities, but is now formulated in the concepts developed in (9.18 to 9.21):

1. A is now a subset of a finite set of all relevant descriptions of the activities DM has found relevant for an attainment of his objectives in an environment E.
2. A is also assumed to describe not only DM's available (=C) and relevant activities, but also identifiable (re-) actions of E and resultant activities co-produced by ιwo or more of DM's activities and/or E's (re-) actions. These options are open to us because of the intra- and interrelations.
3. Due to the intra- and interrelations, the set of activities is functionally a

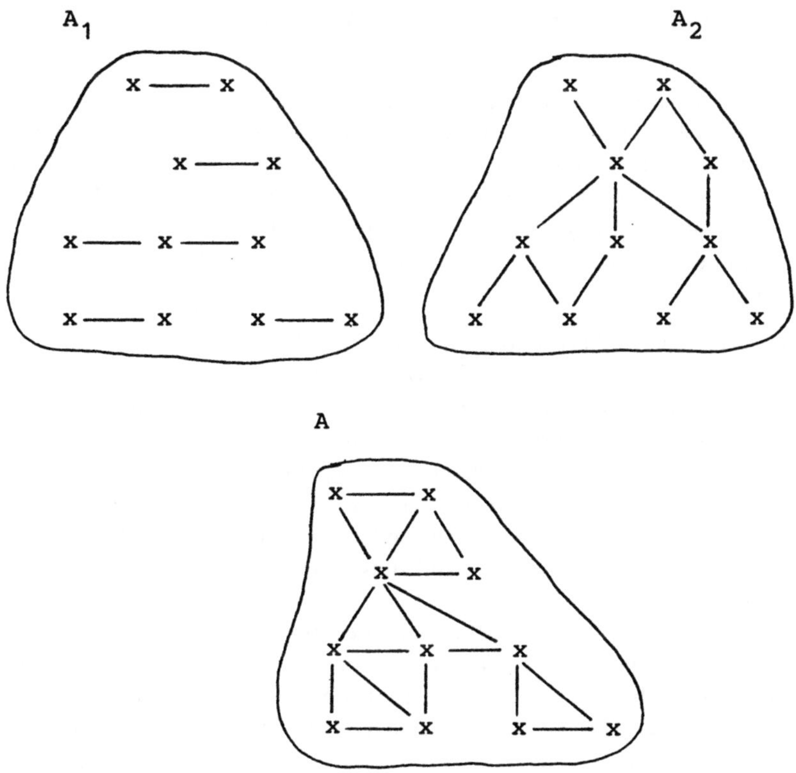

Figure 9.5. The Set $A \subset A_1 \times A_2$

whole, where this whole possesses characteristics other than those of the single activities (for example, the activities of a single transistor versus a transistor radio). But *structurally,* the set is a divisible whole, which makes it possible to describe and analyze the relations between single activities, activities and subsets of activities, between subsets of different magnitudes, and so on.

4. The state of A at some t_1, which is the set of "active" elements of A_1, A_2, B and C at t_1, represents a possible configuration of activities that could take place at the same time.

But of the last two properties, the first one is a key property of a *system,* and the second one could as well describe the *state* of a system. As furthermore, A is a finite set, all the elements of which are joined to at least one other element through an intra- or interrelation, and as the relations represent both transformation and coordination, the set A has all the properties generally ascribed to a system (see Ackoff and Emery, 1972). We will therefore make use of A for a systems definition:

> A *system* is an entity formed by a relation on the set of activities, organized hierarchically through an ordering rule so that there are at least two levels in the set, with at least two activities of the lower level and at least one activity of the higher level, and so that every activity is linked to at least one other activity through a relation. The relations are of two kinds—intrarelations performing "coordination" and interrelations performing "transformation." (9.22)

Also observe that we have two claims of a formal nature on the concept for the present purpose: The elements of a system should be formulated in concepts that are, or could be made, internally consistent; all elements of a level of hierarchy h_i should be transformable to another level of hierarchy $h_j (i \gtrless j)$.

As a consequence of the last claim, a movement $h_i \rightarrow h_j (i < j)$ in a hierarchical system represents an ***aggregation*** of the activities and thus a change to a higher level of abstraction; the corresponding movement when $i > j$ represents a ***disaggregation*** of the descriptions, and thus correspondingly a change to a lower level of abstraction.

The second property of the set of activities (see above) implies that the set of outcomes O described earlier, should be implemented as a subset of A, as it was assumed to describe also resultant activities coproduced by two or more of DM's activities and/or Es (re-)actions. There remains to formulate the sets E and G in the present concepts.

If the set of activities is represented as a system with a hierarchical structure, it would be desirable to formulate the set G in such a way that the objectives could be arranged hierarchically and implemented for activities of different levels of aggregation. For that purpose we will apply the following formulation:

An objective $G_o^{(h)}$, which is a subset of G, is a subset of one or more sets describing activities, i.e., $G_{ko}^{(h)} \subseteq \alpha_{ki}^{(h)}$. (9.23)

A straightforward interpretation of this definition suggests that an objective represents some wanted state of one or more activities—that is, a set of objectives would represent a wanted state of the system, which in turn corresponds to some configuration of activities. As these are activities of DM, resultant activities, and (re-)actions, they could all be implemented as a state of the system of activities. If DM wishes to attain more than one objective with a decision, he could at least describe his multiobjective decision problem in this fashion.

From (9.23) it is found that

1. Activities and objectives would be expressed in compatible units.
2. Objectives can be aggregated and/or disaggregated.
3. Objectives, even of different levels of aggregation, could be made simultaneously operational if it is possible to establish the corresponding, necessary configuration of actions.
4. $G \supset G_{ko}^{(h)}$, $\forall h[\forall o(\forall k)]$, that is, the objectives of G should be internally consistent; theoretically we could, however, implement inconsistent sets of objectives $G_1, G_2, \ldots, G_n$, if we do not try to make them simultaneously operational.

Clearly this seems to be a way of tackling the *multiobjective decision problem.* For modeling purposes this problem is, however, not yet solved with (9.23); we still have to develop an algorithm that would implement (9.23) and also create a program that gives necessary and sufficient conditions for an attainment of all the objectives of G. Figure 9.6 shows a system of the type A with a few objectives implemented. In Figure 9.6, $\alpha_k^{(h)}$ represents multidimensional activities; $G_j^{(h)}$ single- and multidimensional objectives, which are implemented for one or more objectives. Horizontal lines are intrarelations. Vertical arrows are interrelations, and two-directional arrows indicate that both aggregation and disaggregation are possible.

There remains the task of outlining and defining the environment E in the present concepts:

Given a system based on the set A of activities, a set E of elements, which are not included in the system but have an influence on the system's elements and/or are affected by these at some t_1, or in an interval $[t_1, t_2]$, forms the *environment* of the system. (9.24)

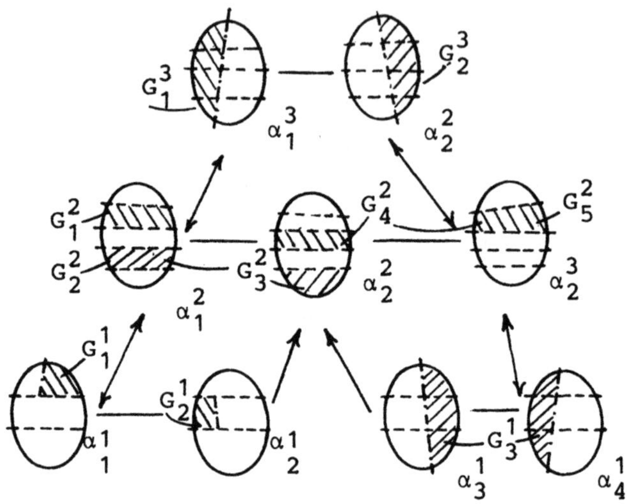

Figure 9.6. A Three-Level System with Ten Implemented Objectives

Let us then apply the terminology developed for A on the elements of E. Let us assume that the elements of E are *activities* and thus represented by a finite set describing all their relevant appearances (to DM) at t_1 or in $[t_1, t_2]$; the state of the environment at t_1, or in $[t_1, t_2]$, is the finite set of all relevant descriptions (in numerical form) of the activities that according to DM form the environment. Interactions between the sets A and E are represented by interrelations because a transition from the poorly outlined elements of an environment to the well-defined concepts forming a system normally will require a transformation.

Through interrelations we can also deal with the old problem of delimiting a system from its environment. The crucial point is to make the boundary both DM and context dependent, so that an activity from time to time can belong to either the system or its environment. Another advantage that can be obtained by using interrelations is that by modifying them we can get an almost continuous transition from closed to open systems and vice versa.

In analogy with our descriptions of the M-situation and the W-problem, Figure 9.7 (in which the elements are simplified from Figure 9.6) describes DM's decision-making situation in the systems concepts we have introduced in this section. In Figure 9.6, x represents an activity, ⓧ is an activity for which an objective is implemented, → is an interrelation, and – an intrarelation. The set G is

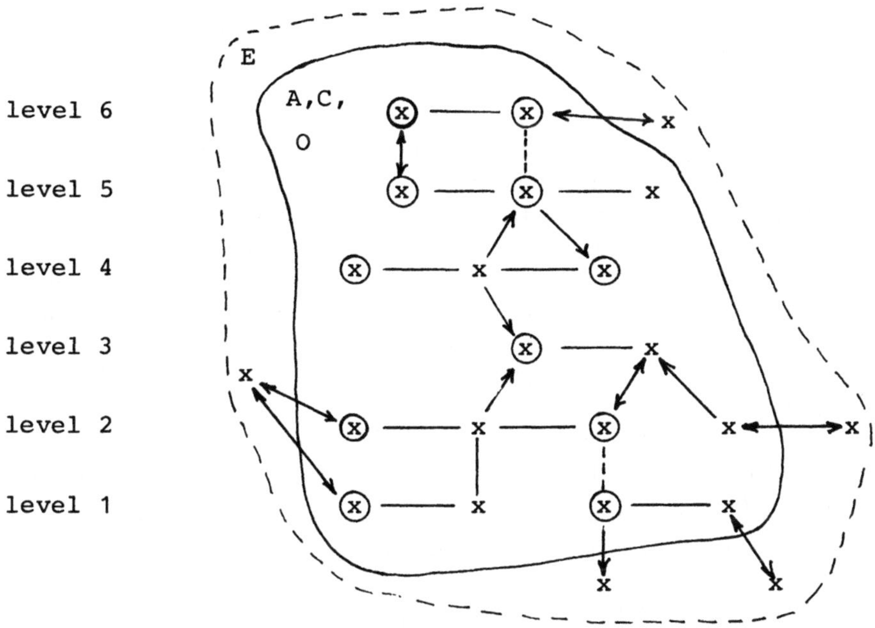

Figure 9.7. DM's Decision-Making Situation

formed by all the activities denoted ⓧ. - - - indicates that an objective is implemented for more than one activity; when vertical and denoted – – →, it indicates that an objective is implemented on more than one level of aggregation.

Now this representation of DM's decision-making situation differs considerably from the *W*-problem:

1. DM's activities and their outcomes, in the *W*-problem represented by the sets *A*, *C* and *O*, but now by the single set *A*, interact through intra- and interrelations; the interaction takes various forms due to the functional characteristics given the relations.
2. Activities, as well as relations, may be defined either for some t_1 or for an interval $[t_1, t_2]$.
3. The objectives are implemented as subsets of one or more activities, and there is no need to define the objectives in terms of the outcomes only [compare $G_1 = \emptyset$ in assumptions giving (9.4) to (9.8)].
4. The environment is described by a set of activities that interact with *A* through interrelations; thus the environment will have different influences on the activities of a system, and one activity may show different in-

fluences in an interval $[t_1, t_2]$ (compare the unit set $\{e'\}$ of the *W*-problem).

Thus the systems approach seems to give a fairly realistic model of DM's decision-making situation. The suggested approach might, in fact, be useful as a means for catching the true complexity of a "mess" (cf. Ackoff, 1974; Carlsson, 1977 *a,b*), which is not very far from the "true complexity of a genuine management problem." But a realistic model implies, more often than not, that we will have difficulties in developing an efficient problem-solving algorithm—and the present case is no exception. Applying the concepts developed for the *W*-problem, a *winning* strategy would be one that creates necessary and sufficient conditions for G as a whole to be attained at t_1, or all the elements of G at least once in an interval $[t_1, t_2]$.

Let us consider the case of *one* activity $a_{ki}^{(h)}$ and *one* objective, implemented for $a_{ki}^{(h)}$, that should be attained at t_1; at $t_{1-\epsilon}$, where ϵ is a short "reaction time," let $a_{ki}^{(h)} \neq G_{ko}^{(h)}$. Then for a winning strategy [see (9.16)] to exist for $a_{ki}^{(h)}$, there should be an integer N such that there exists a sequence

$$[a_{ki}^{(h)}(t_{1-\epsilon}), \beta_{kij}^{(h)}(a_{kj}^{(h)}, t_{1-\epsilon}), C(A^{(\bar{h})}, t_{1-\epsilon})]_1, [\ldots]_2, \ldots, [\ldots]_n, \forall \bar{h}, \qquad (9.25)$$

where $\bar{h} \neq h$, through which we have $a_{ki}^{(h)}(t_1) = G_{ko}^{(h)}$ for $n \leqslant N$.

There are intra- and interrelations in (9.25) such that $a_{kj}^{(h)} = g_1(a_{ki}^{(h)}, \beta_{kij}^{(h)})$, $A^{(\bar{h})} = g_2(a_{ki}^{(h)}, C)$, where g_1, g_2 are auxiliary functions. Then, what is stated in (9.25) is essentially that we, in a winning strategy for $a_{ki}^{(h)}$, will have to consider all intrarelated activities and all interrelated activities, their *effects* on $a_{ki}^{(h)}$ and the *effects* of $a_{ki}^{(h)}$ on them, their *countereffects* on $a_{ki}^{(h)}$ and its *counter-countereffects,* and so on. For two or more objectives, we should add the effects of interdependences among the objectives; when, furthermore, adding one more activity, we will have to consider one more set of interdependences, and so forth. Then the conclusion seems well founded that we will need rather powerful algorithms to be able to realize an attainment of the whole of the set G, in and with the set A of activities.

In the next section we will show how this systems concept might be operationalized and used as a basis for a modeling technique. In Carlsson (1977*a*) an approach using fuzzy automata is outlined, and in Carlsson (1977*b*) some results with locally implemented adaptive control functions, coordinated with heuristic rules, are reported.

Is this approach, then, more relevant as a decision model than (9.16)? At least complexity and multiobjectivity are not avoided, and as found above, the simplifying assumptions necessary for formulating a *W*-problem were not needed.

The systems approach is, in fact, an exponent of a new methodology for complex decision problems. It is a methodology based on *expansionism*—an aspiration to study entities as parts of a larger whole, *synthesis*—a search for ways to combine elements to consistent wholes, and *teleology*—methods for learning and adapting more effectively. In this sense, and as the systems approach has useful and interesting properties as a modeling technique, (9.25) can be said to be more relevant as a decision model than (9.16).

AN OPERATIONALIZATION OF THE SYSTEMS APPROACH

In order to get some experience with the principles outlined above, and in order to test the applicability of the systems approach, an experimental system-simulating model was developed and extensively tested. The model, as shown in Table 9.1, represents a four-level, hierarchical system in which twenty-three objectives of different levels of aggregation are implemented (see Figure 9.8).

An *activity* is represented by the field defined for a variable. An *intrarelation* (—— in Figure 9.8) is represented by the field for a function or a logical relation. An interrelation (——→ in Figure 9.8) also is represented by the field of definition for a function or logical relation. (– – –→ in Figure 9.8) represents a *goalrelation,* which is given the same characteristics as the interrelation but is applied for aggregating/disaggregating objectives. *Objectives* are implemented as subfields of the fields defined for the activities. The units shown in Table 9.2 (~ log norm, ~ norm, etc.) are random number generators with log normal-, normal-, Poisson- and Erlang- distributions and are applied to generate stochastic

Table 9.1. Activities and Goals for Four-Level System

Level	*Activities*	*Objectives*
1	$a_{11}, a_{12}, a_{13}, a_{23}, a_{24}, a_{31}$ $a_{32}, a_{41}, a_{42}, a_{51}, a_{52}, a_{61}$ $a_{62}, a_{63}, a_{64}, a_{65}, a_{66}, a_{72}$ $a_{73}, a_{74}, a_{75}, a_{76}, a_{78}, a_{82}$ a_{93}, a_{102}	$g_{12}, g_{13}, g_{14}, g_{41}, g_{51}, g_{71}, g_{91}, g_{92}$
2	$a_{14}, a_{21}, a_{22}, a_{33}, a_{34}, a_{43}$ $a_{44}, a_{53}, a_{54}, a_{67}, a_{68}, a_{69}$ $a_{71}, a_{77}, a_{81}, a_{91}, a_{101}, a_{8}$	$g_{11}, g_{21}, g_{22}, g_{31}, g_{61}, g_{62}, g_{63}, g_{101}$
3	a_5, a_6, a_7, a_9	g_7
4	a_1, a_2, a_3, a_4	$g_1, g_2, g_3, g_4, g_5, g_6$

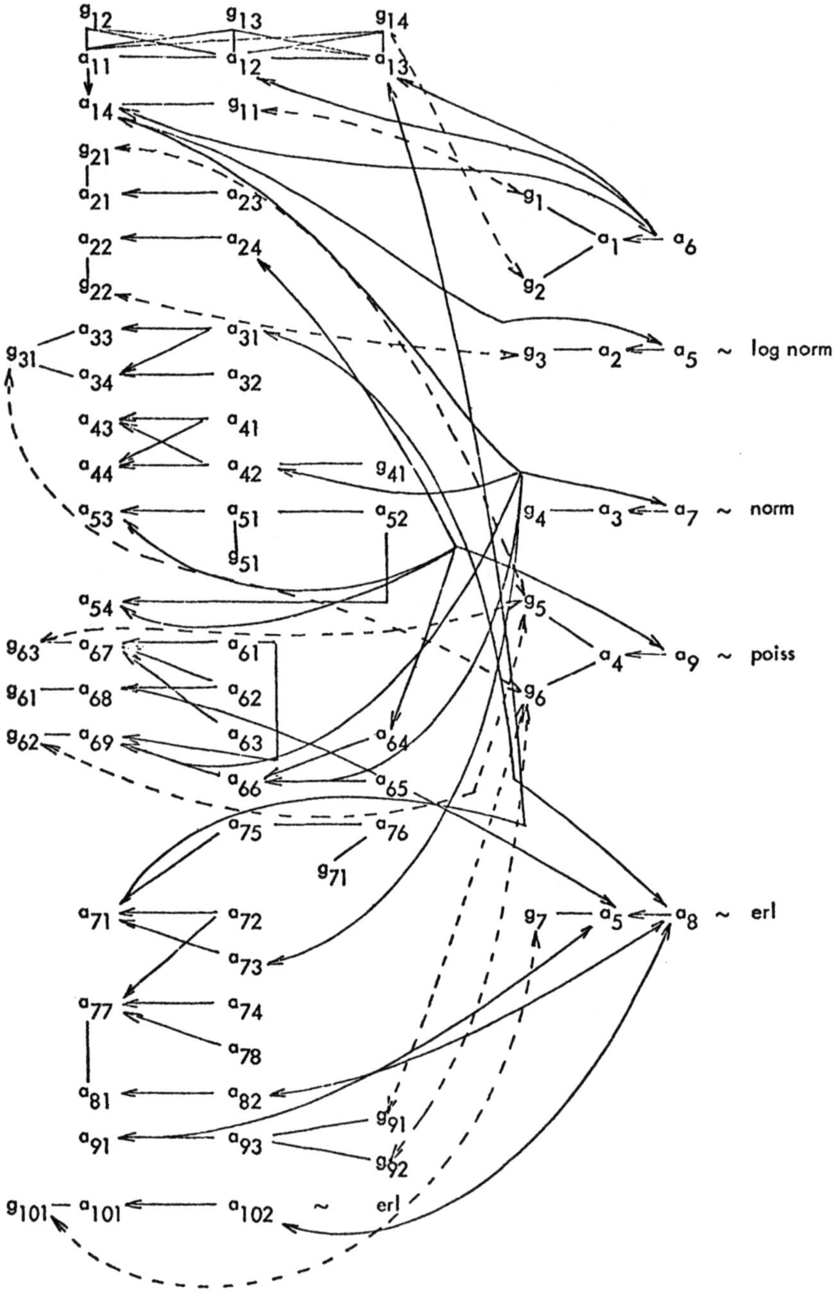

Figure 9.8. B9333, a System-Simulating Model

Table 9.2. Random Number Generators

CTOP				
	a_6	a_1		
	7.50	18.60	g_1	10.00 (⩽)
			g_2	19.70 (⩾)
LOG NORM (0.53, 21.29)	$\sim a_5$	a_2		
		15.57	g_3	10.63 ± 5.49
NORM (12.00, 1.07)	$\sim a_7$	a_3		
		8.32	g_4	13.00 (⩽)
POISS (1.43, 0.38)	$\sim a_9$	a_4		
		4.34	g_5	2.28 (⩾)
			g_6	14.00 (⩽)
ERL (16.00, 0.96)	$\sim a_8$	a_5		
		0.09	g_7	19.99 (⩽)

disturbances in the system. The functional characteristics of the system are shown in Table 9.3 (for more details on B9333, see Carlsson, 1977*a*). In Table 9.3 CTOP, CDS 1 to CDS 10 are subsystems of which CTOP is hierarchically of a higher level than CDS 1 to CDS 10.

How then should the objectives be realized in a systems context of the type B9333 represents? An approach that has been tested is to use locally implemented, adaptive control functions, which are coordinated by a set of heuristic rules. These functions are expected to form a global adaptation process, which in a finite number of steps would bring the system into a state where all the objectives are attained.

Adaptivity implies essentially an ability to react efficiently to changing conditions of both a functional and structural character (cf. Ackoff and Emery, 1972). At its best, a program for adaptive control could be functionally isomorphic with the context in which it is implemented; it would be an "efficient" and/or "purposeful" reaction, with operational means on characteristics or processes in the context essential for the controlled object or system.

The adaptive control functions implemented in B9333 are designed to carry out a program that would be functionally isomorphic, both with "its" subsystem and the whole of B9333:

Table 9.3. Functional Characteristics of B9333

CDS 1						
	a_{11}	a_{12}	a_{13}	a_{14}		
	5.32	4.37	6.08	217.3	g_{11} 217.3 (⩾)	
	4.32	4.37	7.15	217.3		
	3.32	4.37	8.22	217.3	g_{12} 34.0 (⩽)	
	6.12	3.37	6.08	217.3		
	7.92	2.37	6.08	217.3	g_{13} 30.0 (⩽)	
	6.25	4.37	5.08	217.3		
	7.18	4.31	4.08	217.3	g_{14} 78.9 (⩽)	
CDS 2						
	a_{23}	a_{21}	a_{24}	a_{22}		
	6.35	10.19	8.85	8.84	g_{21} 10.20 ± 1.53	
	6.85	11.00	7.85	8.84		
	5.85	9.39	9.85	8.84	g_{22} 8.85 ± 1.32	
	7.30	11.73	6.85	8.84		
	5.40	88.67	10.85	8.84		
CDS 3						
	a_{31}	a_{33}	a_{32}	a_{34}	a_{35}	
	4.00	24.31	6.00	9.72	2.50	$g_{31} = 2.50 \pm 0.37$
	4.13	25.93	5.87	9.70	2.67	
	3.87	22.76	6.13	9.75	2.33	
	4.25	27.45	5.75	9.69	2.83	
	3.75	21.37	6.25	9.77	2.19	$a_{35} = a_{33}/g_{34}$

CDS 4

a_{41}	a_{43}	a_{42}	a_{44}		
3.20	43.23	4.90	0.02	g_{41}	4.89 ± 0.62
3.45	49.66	5.21	0.02		
2.95	37.28	4.59	0.03		
3.70	56.57	5.52	0.01		
2.70	31.79	4.28	0.04		

CDS 5

a_{52}	a_{51}	a_{53}	a_{54}		
0.56	3.88	1.36	1.96	g_{51}	4.28 ± 0.60
0.50	4.12	1.38	1.85		
0.61	3.70	1.35	2.04		
0.44	4.32	1.40	1.74		
0.38	4.51	1.41	1.61		

CDS 6

a_{61}	a_{62}	a_{63}	a_{67}	a_{68}	a_{64}	a_{65}	a_{66}	a_{69}		
0.64	0.96	5.00	25.41	4.23	4.30	8.25	14.19	2.64	g_{61}	4.55 ± 0.68
0.73	1.01	5.50	26.24	4.45	4.80	8.50	15.00	2.40		
0.68	0.91	5.20	24.59	4.13	3.80	8.00	13.40	2.62	g_{62}	2.64 ± 0.39
0.60	1.06	6.00	27.06	4.68	5.30	8.75	15.80	2.64		
0.55	0.86	5.70	22.20	4.10	3.30	7.75	12.60	3.04	g_{63}	23.85 ± 1.65

Table 9.3. *Continued*

CDS 7

a_{76}	a_{75}	a_{72}	a_{73}	a_{71}	a_{74}	a_{78}	a_{77}	
11.00	2000.00	15.00	1420.00	1359.98	0.08	0.02	842.61	g_{71} 11.20 ± 1.68
12.04	2100.00	16.13	1520.00	1001.67	0.18	0.12	904.02	
10.36	1900.00	13.87	1320.00	1160.98	0.09	0.06	1104.79	
12.88	2200.00	17.25	1620.00	879.63	0.28	0.22	965.14	
9.52	1800.00	12.75	1220.00	836.62	0.14	0.11	1018.33	

CDS 8		CDS 9		CDS 10	
a_{82}	a_{81}	a_{93}	a_{91}	ERL ~ a_{102} (20.45, 6.14)	a_{101}
14.00	0.25	1.80	0.25	0.95	−0.53
15.05	0.18	1.63	0.18	0.75	−0.54
12.95	0.27	1.93	0.30	1.15	−0.52
16.10	0.10	2.23	0.41	0.55	−0.55
11.90	0.23	2.53	0.51	1.35	−0.53
		g_{91} 1.78 ± 0.15		g_{101} 22.00 (⩽)	
		g_{92} 2.20 ± 0.33			

1. The incentive for the controller is an "observation" that an objective is not attained; there is at least one adaptive controller surveying each objective.
2. The adaptation process is aimed at modifying the states of those activities for which the objective is implemented, in a way that is "efficient" in relation to the incentive; the adaptation process is governed by parameters determining the length and the number of steps in each run of the process.
3. An adaptive controller operates on free objectives, activities limited by objectives, and intrarelations—in that order—if necessary to carry the adaptation process any further than modifying free objectives; the process stops when the incentive is neutralized or when the process meets a stopping rule, which "hands over" the task to hierarchically higher adaptive controllers.

Applied to the subsystem CDS1 the adaptation process shows the pattern displayed in Figure 9.9. In CDS 1:

1. a_{14} should be $\geqslant g_{11}$; as soon as a_{14} falls below that level an adaptive controller is initiated and raises a_{14} to the wanted level (see Figure 9.9).
2. A "coordination" of a_{11}, a_{12}, a_{13} should be $\leqslant g_{14}$; thus the adaptation process working on a_{14} will also have effects on g_{14} (see Figure 9.9).
3. g_1 and g_2 are "free objectives" in relation to a_1, which should satisfy $g_1 \leqslant a_1 \leqslant g_2$ (the adaptation process can be seen in Figure 9.9); as g_2 and g_{14} interact, the adaptation process operating on g_2 would be superfluous from period 10.

Adaptive controllers operating in a similar fashion are implemented in all other subsystems as well, and as these stand in interaction through interrelations, the controllers will either support or counteract each other. In the first case we get a nice global adaptation process; in the second we get oscillations.

A coordination of the local adaptation processes, aimed at creating a form for global adaptive control, is probably a rather difficult task. In Carlsson (1977*a*) an approach applying heuristic coordination rules was attempted, and the results were not completely discouraging, although not yet definitively positive. But the systems approach outlined here seems to be a fairly powerful technique for dealing with complex decision-making situations involving multiple objectives. Even very messy situations can be dealt with in terms of activities, intra- and interrelations (cf. Carlsson, 1977*b*), and adaptive control has an intuitive appeal as a basis for algorithms dealing with the problem of multiobjectivity. Although not much more than an embryo, we find it a positively promising endeavor to develop this systems approach as a modeling technique.

	a_{11}	a_{12}	a_{13}	a_{14}	a_1	a_6	g_{11}	g_{12}	g_{13}	g_{14}	g_1	g_2
1.	5.32	4.37	6.07	217.21	18.60	7.50	216.00	34.00	30.00	56.00	10.00	19.70
2.	5.32	4.37	6.07	217.21	18.60	7.50	216.00	34.00	30.00	56.00	10.00	20.68
3.	5.32	4.37	6.07	217.21	18.60	7.50	216.00	34.00	30.00	56.00	10.00	21.72
4.	5.32	4.37	6.07	217.21	18.60	7.50	216.00	34.00	30.00	56.00	10.00	22.81
5.	5.32	4.37	6.07	217.21	23.22	9.81	216.00	34.00	30.00	56.00	10.00	23.95
6.	5.32	4.37	6.07	217.21	23.22	9.81	216.00	34.00	30.00	56.00	10.00	20.63
7.	5.32	4.37	6.07	217.21	23.22	9.81	216.00	34.00	30.00	56.00	10.00	21.67
8.	5.32	4.37	6.07	217.21	23.22	9.81	216.00	34.00	30.00	56.00	10.00	23.89
9.	5.32	4.37	6.07	217.21	23.22	9.81	216.00	34.00	30.00	56.00	10.00	22.75
10.	5.32	4.36	6.07	188.86	23.22	9.81	215.89	34.00	30.00	64.74	9.93	23.89
11.	5.85	4.33	6.03	222.87	23.09	9.74	215.89	35.70	31.50	78.70	9.93	23.89
12.	5.85	4.77	6.63	222.87	23.09	9.74	215.89	37.48	33.07	78.70	9.93	23.89
13.	5.85	4.75	6.61	308.98	24.97	10.68	215.89	37.48	33.07	78.70	9.93	25.08
14.	6.44	5.23	7.28	308.98	24.97	10.68	215.89	39.36	34.73	78.70	9.93	25.08
15.	7.08	5.75	8.00	308.98	24.97	10.68	215.89	41.32	36.47	78.70	9.93	25.08
16.	7.78	6.32	8.80	308.98	24.97	10.68	215.89	43.39	38.29	78.70	9.93	25.08
17.	7.78	6.32	8.80	308.98	24.97	10.68	215.89	45.56	40.20	82.63	9.93	26.34
18.	7.78	6.32	8.80	308.98	24.97	10.68	215.89	45.56	40.20	82.63	9.93	27.66
19.	7.78	6.32	8.80	308.98	24.97	12.78	215.89	45.56	40.20	82.63	9.93	29.04
20.	7.78	6.32	8.80	308.98	24.97	12.78	215.89	45.56	40.20	82.63	9.93	30.44
21.	7.78	6.32	8.80	308.98	32.01	14.20	215.89	45.56	40.20	82.63	9.93	32.01
22.	7.78	6.32	8.79	205.67	32.01	14.20	215.89	45.56	40.20	95.54	9.93	30.45
23.	7.78	6.20	8.63	295.03	31.50	13.95	215.79	45.56	40.20	105.33	9.93	31.97
24.	7.78	6.19	8.61	222.68	31.43	13.91	215.69	45.56	40.20	121.78	9.93	31.97
25.	7.78	6.19	8.61	222.68	31.03	13.71	215.69	45.56	40.20	121.73	9.92	36.97
26.	7.78	6.10	8.49	290.21	31.03	13.71	215.59	45.56	40.20	140.81	9.92	42.74
27.	7.78	6.10	8.49	290.21	30.97	13.68	215.59	45.56	40.20	140.81	9.92	42.74
28.	7.78	6.08	8.47	215.03	30.97	13.68	215.49	45.56	40.20	162.80	9.91	49.42
29.	7.78	5.99	8.34	226.38	30.55	13.47	215.39	45.56	40.20	188.23	9.91	57.14
30.	7.78	5.99	8.34	226.38	30.21	13.30	215.39	45.56	40.20	188.23	9.91	57.14
31.	7.78	5.92	8.23	197.81	30.21	13.30	215.29	45.56	40.20	217.63	9.90	66.06
32.	7.78	5.92	8.23	197.81	29.76	13.08	215.29	45.56	40.20	217.63	9.90	66.06
33.	7.78	5.82	8.09	193.60	29.76	13.08	215.29	45.56	40.20	251.62	9.90	76.38
34.	7.78	5.82	8.09	193.60	29.30	12.85	215.29	45.56	40.20	251.62	9.90	76.38
35.	7.78	5.72	7.95	218.13	29.30	12.85	215.09	45.56	40.20	290.92	9.89	88.32
36.	7.78	5.72	7.95	218.13	28.99	12.69	215.09	45.56	40.20	290.92	9.89	88.32
37.	8.57	6.22	8.65	218.13	28.99	12.69	215.09	45.56	40.20	290.92	9.89	102.11
38.	9.42	6.83	9.51	328.31	34.27	15.33	214.99	45.56	40.20	336.37	9.89	102.11
39.	9.42	6.82	9.49	186.41	34.27	15.33	214.89	45.56	40.20	388.90	9.88	118.06
40.	9.42	6.82	9.49	186.41	33.55	14.97	214.89	45.56	40.20	388.90	9.88	118.06
41.	9.42	6.66	9.27	243.17	33.55	14.97	214.79	45.56	40.20	449.65	9.88	136.50
42.	9.42	6.66	9.27	243.17	33.13	14.76	214.79	45.56	40.20	449.65	9.88	136.50
43.	9.42	6.57	9.14	190.73	33.13	14.76	214.69	45.56	40.20	519.88	9.88	157.82
44.	9.42	6.57	9.14	190.73	32.49	14.44	214.69	45.56	40.20	519.88	9.88	157.82
45.	9.42	6.42	8.94	271.90	32.49	14.44	214.59	45.56	40.20	601.04	9.87	182.42

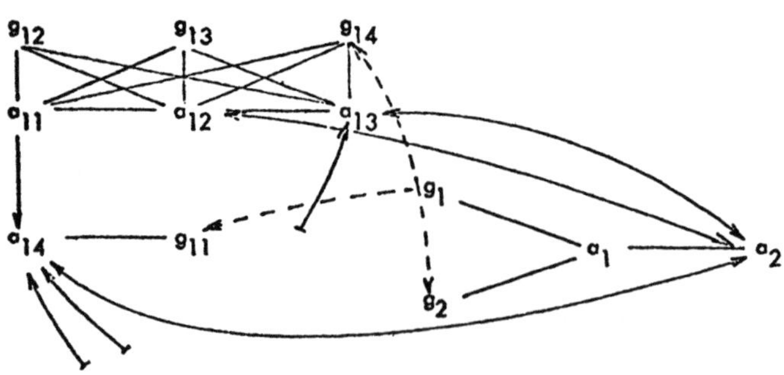

Figure 9.9. CDS 1 and Its Adaptation Process

SUMMARY AND CONCLUSION

We first introduced the modelbuilder's dilemma. Next we studied *M*-situations and *W*-problems in order to find out if complex, multiobjective problems could be described efficiently in that terminology. Then a systems concept was introduced and its impact on and relevance for multiobjective decision models was discussed in some detail. Finally, an operationalization of that concept was outlined and illustrated with some numerical material. To conclude, we found that

1. The concepts introduced by *M*-situations and *W*-problems are not appropriate for dealing with complex, multiobjective problems.
2. The systems approach introduced with our systems concept seems well suited for dealing with problems associated with "complexity" and "multiobjectivity."
3. Our systems approach has some interesting and promising features as a basis for a modeling technique: more than one objective can be implemented simultaneously, activities of different levels of abstraction can be described simultaneously, and a model can be dynamic and/or static.

All of the above make the suggested approach an interesting alternative to traditional research efforts on multiobjective decision models.

REFERENCES

Ackoff, R.L. 1974. "Beyond Problem Solving." *General Systems* 19.

Ackoff, R.L., and F.E. Emery. 1972. *On Purposeful Systems.* Chicago: Aldine/Atherton.

Banerji, R.B. 1969. *Theory of Problem-Solving: An Approach to Artificial Intelligence.* New York: American Elsevier.

Bunge, M. 1969. "The Metaphysics, Epistemology and Methodology of Levels." In L. Whyte et al., eds. *Hierarchical Structures.* New York: American Elsevier.

Carlsson, C. 1977*a*. "On Adaptive Multigoal Control. On the Principles for Operational Problem Solving in a Complex Environment." Ph.D. dissertation (in Swedish), Åbo Swedish University School of Economics, A:15, Åbo.

——. 1977*b*. "An Approach to Adaptive Multigoal Control Using Fuzzy Automata." *Proceedings of EURO II.* Amsterdam: North-Holland.

Kaimann, R.A. 1974. "Coefficient of Network Complexity." *Management Science* 21.

May, R.M. 1972. "Will a Large Complex System Be Stable?" *Nature.*

Mesarovic, M.D. 1972. "Conceptual Foundations for a Mathematical Theory of General Systems." *Kybernetes* 1.

Mesarovic, M.D., et al. 1969. "Foundations for a Scientific Theory of Hierarchical Systems." In L. White et al., eds. *Hierarchical Structures*. New York: American Elsevier.

Newell, A., et al. 1958. "Chess-Playing Programs and the Problem of Complexity." *IBM Journal of Research and Development* 2.

Sahal, D. 1976. "Elements of an Emerging Theory of Complexity per Se." *Cybernetica* 1.

NAME INDEX

Abelson, R. P., 114, 115, 116, 140
Ackoff, R., 6, 16, 166, 170, 174, 177, 183
Alexander, J. M., 109, 110
Arib, M., 20, 24
Aronson, W. J., 115, 116, 140
Assilian, S., 30, 32, 39
Athy, K. R., 149, 156
Atkin, R., 8, 16, 51, 60, 67, 73, 74
Attleslander, P., 154, 156
Azar, E. E., 76, 92

Back, K. W., 154, 156
Banerji, R. B., 162, 165, 183
Banks, A. S., 75, 77, 78, 85, 92
Barto, A., 78, 92
Bellman, R. E., 29, 34, 39
Ben-Dak, J., 76, 92
Berlinski, D., 26, 39
Bernstein, A., 114, 116, 140
Birthwistle, G., 116, 117, 140
Bishop, V. F., 75, 92
Bourbaki, 18
Bradburn, N. M., 150, 154, 157
Bråten, S., 114, 116, 120
Brinton, C. C., 84, 92
Broekstra, G., 88, 92
Bryant, E. C., 149, 156
Bunge, M., 168, 183

Cannell, C. F., 154, 156
Carlsson, C., 174, 177, 181, 183
Carter, G.A., 30, 40
Cartwright, D., 116, 141
Cavallo, R. E., 5, 7, 8, 12, 13, 14, 16, 17, 19, 22, 24, 43, 75, 77, 79, 87, 88, 89, 90, 92, 140, 141
Clifford, 53, 54
Coleman, J. E., 114, 141, 149, 156
Cortés, F., 7, 17

Dahl, O. J., 116, 117, 126, 140, 141

Davies, J.C., 84, 92
Davis, P., 78, 92
Descartes, 3
Deutsch, K. W., 76, 92
Dijkstra, W., 143, 154, 156
Downes, B., 84, 92
Dubois, D., 27, 29, 30, 39

Edwards, L., 84, 92
Einstein, 53, 54

Feierabend, I., 85, 92
Feierabend, R., 85, 92
Firestone, J., 75, 92

Gaines, B. R., 8, 17
Gantmacher, F. R., 97, 110
Gardner, I., 156
Gergen, K. J., 154, 156
Getzels, J. W., 154, 156
Goguen, J. A., 20, 24
Goldman, M., 156
Gorden, R. L., 154, 156
Green, J. J., 154, 157
Guetzkow, H., 116, 117, 141
Gurr, T. R., 75, 84, 92

Harary, F., 116, 141
Hayek, F. A., 4, 17, 91, 92
Heisenberg, W., 3, 4, 5, 6, 17, 19
Hibbs, D. A., 75, 85, 93
Hoos, I., 26, 39
Huntington, S. P., 87, 93

Jahren, E., 114, 116, 141
Jain, R., 27, 29, 30, 39

Kahn, R. L., 154, 156
Kaimann, R. A., 161, 183
Katz, E., 114, 141
Kaufmann, A., 33, 39
Khouja, M. W., 100, 110
Kickert, W. J. M., 30, 40
King, P. J., 30, 40
Klir, G. J., 8, 9, 12, 13, 17, 19, 22, 24, 87, 88, 89, 92, 93, 115, 116, 129, 130, 133, 136, 140, 141
Kneubühler, H. U., 154, 156
Knuth, D. E., 28, 40
Kolson, K. L., 154, 157
Kornhauser, W., 84, 93
Kotler, P., 116, 117, 141

Lazarsfeld, R., 7, 17, 114, 141
Lewis, P. M., 28, 40
Löfgren, L., 5, 17
Lupsha, P., 84, 93

Macko, D., 138, 141
Mamdani, E. H., 30, 32, 40
Manes, E. G., 20, 24
May, R. M., 161, 183
McCormick, D., 75, 92
McCulloch, W. S., 140, 141
McGuire, W. J., 115, 116, 140
Menzel, H., 114, 141
Merton, R. K., 116, 141
Mesarovic, M. D., 8, 17, 22, 23, 24, 114, 138, 141, 168, 184
Miller, G. A., 98, 110
Mizumoto, M., 27, 29, 40
Moore, R. E., 29, 40
Myrhaug, B., 116, 117, 140

Næss, A., 116, 141
Nahmias, S., 27, 29, 40
Nauta Lemke, H.R. van, 30, 40
Negoita, C. V., 26, 33, 40
Newcomb, T. M., 115, 116, 140
Newell, A., 161, 184
Newton, I., 54
Norlén, U., 116, 141
Nowakowska, M., 153, 154, 157
Nygaard, K., 116, 117, 126, 140, 141

Ostergaard, J. -J., 30, 40

Pareto, V., 91, 93
Pask, G., 8, 17

Pestel, E., 22, 24
Pettee, G., 84, 93
Piaget, J., 8, 17
Pichler, F., 8, 17, 22, 24
Popper, K. R., 7, 17
Prade, H., 27, 29, 30, 39
Przeworski, A., 7, 17

Ralescu, D. A., 26, 33, 40
Reitman, A. P., 149, 156
Rescher, N., 7, 8, 17
Rogers, E., 114, 141
Rogers G., 82, 93
Rosenberg, M. J., 115, 116, 140
Rosenkrantz, D. J., 28, 40
Rummel, R. J., 75, 76, 77, 78, 85, 90, 93
Russell, B., 47, 48
Rutherford, D., 30, 40

Saaty, T. L., 97, 98, 100, 109, 110
Sahal, D., 161, 184
Sanchez, E., 32, 40
Schultz, R., 116, 117, 141
Singer, J. D., 9, 10, 12, 17, 76, 93
Spiegel, J. P., 84, 93
Sprague, J., 7, 17
Stearns, R. E., 28, 40
Sudman, S., 150, 154, 157
Summers, G. F., 154, 157
Svenning, L., 114, 141

Takahara, Y., 8, 17, 22, 23, 24, 138, 141
Tanaka, K., 27, 29, 40
Tang, J., 149, 156
Tannerbaum, P. H., 115, 116, 140
Tanter, R., 75, 84, 85, 93
Taveggia, T. C., 149, 157
Terrell, L. M., 85, 93
Teune, H., 7, 17
Tong, R. M., 30, 32, 40

Uyttenhove, H., 13, 17, 88, 93

Valach, M., 115, 141

Wenstøp, F., 30, 32, 41
Williams, J. A., 154, 157
Winch, P., 20, 24
Winston, P. H., 30, 41
Wymore, A. W., 8, 17, 19, 24

Zadeh, L. A., 5, 25-36, 38, 39, 41
Zeigler, B. P., 8, 17, 129, 130, 141
Zouwen, J. van der, 143, 154, 156, 157

SUBJECT INDEX

Abstract languages, 43
Abstraction, 1, 168
Accuracy prediction, 147
Activities, 159, 163, 167, 171, 175
Activity units, 158
Actors, 144
Adaptive control functions, 174
Aggregation/disaggregation, 169, 170, 175
Algebra, 18; exterior, 67
Algorithms, 12, 83, 88, 174
Analysis, 18, 166; black box, 142; correlation, 76; factor, 76, 86, 91, 153; regression, 69, 76
Anomalous states, 82, 86
Anomaly, 90
Appearances, 11
Approximation, 34, 90
Artificial intelligence, 30
Assumptions, 9, 16, 144
Attitudinal data, 152
Attributed grammar, 28
Attributes, 11, 78, 79; basic, 11, 78; supporting, 11, 79
Auxiliary hypotheses, 142, 144

Backcloth, static, 50, 53, 60
Basic attributes, 11, 78
Behavior, 9, 11, 143, 165
Behavior systems, 11, 81, 86
Behavioral scientists, 159
Black box analysis, 142, 143
Boundary, 172

Categories, theory of, 20
Cause-effect, 166
Clustering, 100
Cognitive consistency, 114-119, 132
Communication network, 115
Complex systems, 6, 8
Complexity, 1, 4, 5, 7, 15, 19, 26, 90, 113, 159, 161, 174

Composite terms, 28
Computer science, 43
Conceptual framework, 5, 26, 158
Conceptual model, 142
Conflict: domestic, 84, 86; within nations, 75; resolution of, 109
Connectivity, 54
Consensus-competition models, 82, 83
Consistency, 96, 98
Constructing an algorithm, 158
Context, 43, 78, 82
Control strategies, 32, 164
Coordinating and/or aggregating objectives, 158
Coordination, 170
Correlation analysis, 174
Correlational coefficient, 69, 150
Correlational knowledge, 9
Correlations, pairwise, 88
Countereffects, 174
Covariance prediction, 147
Cover set, 13, 48

Data, 45, 46, 75, 76, 97, 149; attitudinal, 152; empirical, 22; hard, 46; interview, 142; soft, 45
Data: collection, 46, 65, 69, 150; -gathering process, 137; sets, hierarchy of, 45; source, 113; systems, 11, 15, 79, 81
Decision-context, 160
Decision making, 84, 158, 159; process of, 163; strategy of, 162
Decision model, 166
Decompositions, 12, 20, 96
Description, 5, 6, 168
Diagnostics, 32
Domestic conflict, 84, 86
Dyadic balance, 126
Dynamical descriptions, 11
Dynamical hierarchy, 66
Dynamical system, 11, 22, 23
Dynamics, 45, 87, 159

Eccentricity, 56, 58
Econometric models, 22
Economics, 21, 22
Effects, 174
Elite circulation, 91
Empirical data, 22
Environment, 11, 159, 171
Epistemological levels, 9, 14
Existential knowledge, 9
Expansionism, 175
Explanations, 12
Exterior algebra, 67

Factor analysis, 12, 76, 86, 91, 153
Formalism, 162
Framework: conceptual, 5, 26, 158; operational, 6
Fuzziness, 26
Fuzzy: automata, 174; logic 29, 34; logic controllers, 30; number, 27, 29; propositions, 27; sets, primary, 27; systems theory, 5, 26, 34

General methodological approach, 91
Generative systems, 11
Geography, 46
Goalrelation, 175
Goals, 158
GSPS, 6, 8, 11, 81, 87

Heuristic rules, 174
Hierarchical levels, 168
Hierarchical structure, 158, 168, 170
Hierarchy, 9, 14, 48, 87, 94, 103; dynamical, 66; of admissible structured sets, 23; of data sets, 45; of realization methods, 23
Hypotheses, 142, auxiliary, 142, 144

Ill-structured environment, 159
Imprecision, 26, 38
Incidence matrix, 48
Industrial process control, 30
Influence, 100
Information, 12, 14
Information processing, 8, 15, 115
Initial systems, 14, 81
Interaction, 8

Interdisciplinary approaches, 43
Internal states, 124
Interplay, 4
Interpretation, 1, 5, 6, 78, 81, 87
Interrelations, 158, 168, 175
Interview, 142–156
Interview: data, 142; instructions, 142; model, 152, 154; panel, 150; process, 152
Interviewer: effects of, 152; race of, 149
Intransitivity in judgments, 96
Intrarelations, 158, 168, 175
Intuition-amplification, 8
Invariance, shape, 29
Investigator, 8

Judgments, intransitivity in, 96

Knowledge: correlational, 9; existential, 9; explanatory, 9; theory of, 7

Language, 4, 5, 7, 15, 19, 27, 37, 43
Large-scale systems, 25
Lattice structure, 13
Laws, 23, 140
Linear regression, 11, 21
Linguistic fuzzy-relational representation, 30
Linguistic truth-values, 35
Linguistic variable, 27, 33

Management: context, 159; problem, 161; science, 159, 166
Masks, 136
Mathematical relations, 45
Mathematicians, 159
Mathematics, 18, 43
Measurement, 6, 11, 46, 98, 109
Mechanization, 166
Memory, 87
Mess, 174
Metasystem, 13
Methodological, 45; approach, general, 91; procedure, 78, 82; systems, 82
Methodologies: structured, 21; unstructured, 19, 21
Methodology, 7, 20
Model, 159; conceptual, 142
Model: builders, 158; building, 20; users, 158; verification, 15
Modeling, 6, 18, 22, 75
Modeling activity, 6, 8, 87, 111
Multiobjective decision situations, 111, 171
Multiobjectivity problem, 166
Multiplicity of objectives, 159
MULTISIM, 129, 133, 146, 138

Nature, 3
Network: balance, 132; communication, 115; panel, 114, 129
Networks, 107

Object, 9, 11
Object-system, 81
Objectives, 159, 170, 175
Observable variables, 153
Observer, 7
Operations research, 159, 166
Optimization, 104
Organization theory, 30
Output variable, 143

Panel interview, 150
Partition, 48
Pattern, 51, 52, 81
Pattern classification, 32
Physics, 3, 18
Policy, 84, 86, 109
Possibility distribution, 35
Pragmatist, 15
Precision, 38
Prediction, 146; accuracy, 147; covariance, 147; stability, 147; testable, 146
Primary fuzzy sets, 27
Priorities, 94, 100
Priority vector, 105
Problem solving, 158, 159, 163

Problem-solving: behavior, 165; tool, 88
Problems, 8, 43, 159; systems, 14; kinds of, 14
Process, 1, 15, 44
Process character of modeling activity, 87
Psychosocial behavior, 113
"Purposeful" reaction, 177

Questionnaire, 153

Race (of interviewer), 149
Ratio scale, 95
(Re-) actions, 171
Realization methods, 23
Reductionism, 6, 166
Referendum, 134
Regression, linear, 11, 21
Regression analysis, 69, 76
Response function, 143, 144
Revolution, 84
Rule-based systems, 30

School selection, 105
Science, 3
Scientific: data, 45; explanation, 4
Shape invariance, 29
SIMCOM, 114, 117, 138
Simplicial complex, 49
SIMULA, 116, 119, 139
Simulation, 111, 113, 114, 116, 139
Social: disorganization, 86; network, 113; science research, 76; sciences, 7, 19; scientists, 22, 46; structure, 116; systems, 4, 22; systems modeling, 18, 22
Sociological interview, 111
Sociology, 46
Source systems, 11, 78
Stability prediction, 147
State, 9, 11, 168, 170; anomalous, 82, 86; internal, 124; of the system, 87, 171
State: -defined systems, 11; transition function, 143, 144; transition matrix, 128; transition structure, 133; variables, 118, 125, 143, 146
Static backcloth, 50, 53
Strategy: control, 164; decision-making, 162; winning, 164, 174
Structural: backcloth, 60; forces, 51
Structure, 9, 45, 65, 71, 85, 86, 123
Structure: coefficient, 72; systems, 11, 14, 77, 87, 88, 90; vector, 50
Structured methodologies, 21
Symbolic production, 1
Synthesis, 175
System, 168, 170
System: behavior 115; perspective, 114; type, 14
Systems: behavior, 11, 81, 86; data, 11, 15, 79, 81; generative, 11; humanistic, 5, 25, 26; initial, 14, 81; large-scale, 25; object, 81; rule-based, 20; social, 4; source, 11, 78; state-defined, 11; structure, 11, 14, 77, 87, 88, 90; terminal, 14, 81
Systems: -methodological orientation, 82; methodology, 4, 5, 11, 116; problems, 14; research, 4; theoreticians, 159; theory, 5, 25

Teleology, 175
Terminal systems, 14, 81
Testable prediction, 146
Theory: of categories, 20; of knowledge, 7; unification of, 138
Theory/data interface, 7
Time, 137
Tools, 1, 8, 29, 44, 76, 82, 88, 111
Topology, 18
Traffic, 60
Transformation, 170

Uncertain environment, 159
Uncertainty, 159
Unobservable variables, 153
Unstructured methodologies, 19, 21

Winning strategy, 164, 174

LIST OF CONTRIBUTORS

Ronald H. Atkin, Department of Mathematics, University of Essex, Colchester, Essex, England

Stein Bråten, Institute of Sociology, University of Oslo, Oslo, Norway

Christer Carlsson, Institute of Management Science, Åbo Swedish University, School of Economics, Åbo, Finland

Roger Cavallo, Department of Computer and Information Sciences, State University of New York, College of Technology, Utica, New York, U. S. A.

Eivind Jahren, Institute of Sociology, University of Oslo, Oslo, Norway

Arild Jansen, Institute of Sociology, University of Oslo, Oslo, Norway

Franz Pichler, Institute for System Science, Johannes Kepler University, Linz, Austria

Thomas L. Saaty, Graduate School of Business, University of Pittsburgh, Pittsburgh, Pennsylvania, U. S. A.

Lofti A. Zadeh, Computer Science Division, Department of Electrical Engineering and Computer Sciences, and the Electronic Research Laboratory, University of California, Berkeley, California, U. S. A.

Eduard Ziegenhagen, Department of Political Science and Center for Social Analysis, State University of New York, Binghamton, New York, U.S.A.

Johannes van der Zouwen, Department of Research Methodology, Free University, Amsterdam, The Netherlands

MIX
Papier aus verantwortungsvollen Quellen
Paper from responsible sources
FSC® C105338

If you have any concerns about our products,
you can contact us on
ProductSafety@springernature.com

In case Publisher is established outside the EU,
the EU authorized representative is:
Springer Nature Customer Service Center GmbH
Europaplatz 3, 69115 Heidelberg, Germany

Printed by Libri Plureos GmbH
in Hamburg, Germany